LIFE HUNT

BY NEILA SKINNER PETRICK
WITH BARRY ANDREWS CHAMBERS

LIFE HUNT

By Neila Skinner Petrick
with Barry Andrews Chambers

Based on a story by Ivy Opdyke, Barry Andrews Chambers
& Neila Skinner Petrick

Cover Art by Evan Hublein
Copyright © 2013 Neila Skinner Petrick

ISBN-13: 9781492745242
ISBN-10: 1492745243
Library of Congress Control Number: 2013917153
CreateSpace Independent Publishing Platform
North Charleston, South Carolina

DEDICATION

To all who love the earth enough to be good shepherds, even when the cost may be great.

PART I

PART II

PROLOGUE

In the year 2042, there was devastation on the earth caused by man's overreaching greed. It began as a war of seeds. Unaware of the long term consequences, a consortium of nations trying first to enhance and then to control the world's food supply inadvertently created a deadly sub-virus that evolved into a killing machine, swiftly destroying a major life source, seed corn. That was soon followed by the infection of other basic crops. The lack of biodiversity doomed most efforts to save the food supply.

One September morning, unexpected as an alien attack, the contagion exploded, following the sun, cutting a swathe of destruction around the globe. The trees remained, most of them, and the oceans were still the myriad shades of blue. What wasn't easily visible were the faces of the humans which had begun to age at an alarming rate.

The virus-like entity, the unexpected fruit of genetic engineering, had mysteriously entered the food chain. Soon, almost every living thing, plants, animals, humans, was infected and died so rapidly that neither governments nor corporations were able to take countermeasures. When the virus had completed its terrible work, little was left alive. A few resilient people remained, on the aging International Space Station, on Polaris, the shining center of exploration of outer space as well as advanced research in a number of fields, the Antarctic, in a cavern in Norway. Another isolated group remained in the Northeast, where native cranberries were grown in bogs, but they too fell silent. No one knew if they had been stricken with the "plague." If the survivors could find one another in time, the planet's renewal was possible. The work belonged to those who were left. I know the story well. It is my family's chronicle, and I want to share it with you.

Samuel III

PART I

INTRODUCTION

He kept a glass of bourbon on the sink next to his razor blades. Harry Rose looked into the bathroom mirror and saw a stocky, balding man with the kind of bags under his eyes that could be in a commercial for Samsonite Luggage. The bags were from lack of sleep. The deepening lines on his forehead were the results of running a nonstop always demanding business, if you could describe a half-dozen grocery stores that way.

He touched a deep new wrinkle on his chin. "I look like my grandfather," he explained to the image in the mirror, but then turned away.

How did he get so old? He thought he knew. It was the constant struggle of making a payroll and fighting the never-ending demand for new and better products, in addition to dealing with Kenny. Why did he listen to Kenny anyway? Why had he gone into business with that loser? Easy answer. Because Kenny was his wife's younger brother. Fast talking, super salesman Kenny who'd recently added a line of booze and hadn't even bothered to discuss it with his partner. Probably another money loser.

Harry took a long sip of bourbon. His shoulders ached. His legs felt like concrete blocks. Gotta start a jogging program. I am a heart attack waiting to happen. He held the bourbon in a toast up to the older man in the mirror. "To the least changed."

The least changed. That was the award he won at his 20th high school reunion, ten years earlier. Harry was a well-preserved thirty-eight year old then. Other than a little more waistline, his smooth face and jet black hair had amazed his peers.

"Looks like you found the Fountain of Youth," said his old high school buddy, Sonny Ralston. Sonny had been chosen as "Most Handsome" his senior year at Rice High. He still looked pretty good himself with the exception of a moderately receding hairline and the normal wrinkles of any thirty-eight year old. Everyone at the reunion wore their senior picture on the breast pocket to identify themselves to former classmates.

Henry was the hit of the event and he noticed right away that many of the women looked better than they did twenty years before. Maturity had added beauty. Plus good skin care and cosmetics. A few had turned matronly, but you couldn't outrun genetics. Most of the men had paunches, were balding or sported a lot of salt and pepper beards.

Bree Cummings had approached Harry. He glanced surreptitiously at her senior picture to see who she was. The photo revealed a lithe, bleach blond cheerleader. Bree, he thought to himself. Sexy Bree who had filled various fantasies through his high school years. Hell, she was every guy's dream girl. Besides being head cheerleader, she was also captain of the gymnastics team. Harry stared at page one hundred and fifteen of the Rice Panther Roar where there was a picture of Bree on the balance beam. Whew!

Bree waved a chubby hand in Harry's face. "You look just like you did the night we graduated. I don't even know why you're wearing your senior picture. Take off that picture right now, you don't need it."

"Well, you look pretty much the same too, Bree."

She laughed and gave him a playful slap on the shoulder. "Maybe if I took off forty pounds and twenty years." Bree moved in closer and suddenly gave him some serious bedroom eyes. "What's your secret, Harry?" Her voice was suddenly low and husky.

"Good genes and clean living," he joked.

Her hand was holding his and her eyes would not let him go. "I always liked you, Harry. I had a secret crush on you at Rice. Did you know that?"

Harry looked over at his wife who was reminiscing with two old girlfriends. Bree was subtly pressing her groin into his hip.

"You know, Harry, my husband is in Chicago at a convention. I live out around the lake. We could go there for a drink."

Harry remembered their school days when Bree didn't know he existed. She was untouchable, unapproachable. She didn't give him a second glance in the hallway.

"It's tempting, Bree, but my wife probably wouldn't like it."

Her face fell. She suddenly looked even older. Sad. Then she recovered her old cheerleader smile. "Well, it's been great seeing you, Harry."

It was a good memory as Harry took one more swallow of bourbon. At thirty-eight, he was voted "Least Changed" since high school. Now, all he saw was an old man in the mirror. Tomorrow was his 48th birthday.

ONE

Lost In The Stars

POLARIS: OCTOBER 29th, 2042.

The massive space station orbited the earth 15.7 times per day at a speed of more than twenty-seven thousand miles per hour, a rate hardly noticeable to the inhabitants. The earthbound could see it now and then. It was hundreds of miles higher than the orbit of the old International Space Station which still flew its route more than two hundred miles above the earth, but was rarely used except for module storage and space flight training.

Polaris was a mile plus in diameter, composed of multiple concentric rings consisting of labs, warehouses, factories, greenhouses and living areas, and the amazing structure could be seen from the earth without the aid of a telescope. To the naked eye, it appeared as a bright light in the sky.

Polaris had been years in construction. It was a self-supporting system with a population fluctuating between two to six thousand souls from every nation on earth. Someone had facetiously suggested that the moribund United Nations meet there, but the diplomats hated to give up their New York City privileges, particularly the limousines and free parking.

There were stores, entertainment centers and medical facilities all connected by a tube which contained a system of multiple bullet train routes carrying personnel and freight throughout Polaris. Many took pride in beating the host country out of anything they could.

The original purpose was to be the intermediate point for the colonization of Mars and then to seek out other potential sites for outposts in the universe. The mission was slowly changing with the events that had been happening on earth the last 18 months. Commander Jake Addington was called the unofficial "mayor" of Polaris, which was mostly military with a small number of civilians. In his early fifties, he looked ten years younger with the handsome features of a movie star. He had steel gray eyes that seemed to look through the soul of all who came into his sphere. His hair was wavy caramel that always seemed to have a tousled look. The only signs of age were gray streaks mainly at the hair part. Despite the casual features, he was all military and exuded authority the instant he entered a room.

Right now, his features were a mixture of puzzlement and worry as he watched the aero-screen on his computer. He glanced around, looking for his second in command.

"Commander Morris?"

Morris, a wiry little African-American whose family had immigrated to America from Botswana, was at his console and spoke over his shoulder. "I saw it, sir."

"The time phase is correct?" Jake asked.

"I went through the procedures three times. The aero-screen does not lie," Morris said.

"Could it be a satellite transmission malfunction?"

"That was the first thing I checked."

Jake shook his head in disbelief. "Why didn't you call me?"

Morris turned in his chair and gave his boss a sober look. "Is there really anything we can do about it? Besides, I knew you'd be in early. I thought you should get as much sleep as you could."

Jake nodded, "Yeah. You're right. All we can do is monitor at this point. Any word from NASA?"

"No, sir. Every once in a while we'll get something from dispatch. The systems are falling apart down there."

"Yeah," Jake said. "I know."

Instead of feeling helpless, Jake went over his options. There was plenty of grain in storage. The question was, should he send a team? Would they be able to get back? Would they be any help once they got to Houston? It was all unknown.

"Morris, check Program Six."

Morris' eyebrows shot up. "The President's plan?"

"It's my plan, but he approved it. Run it thorough the system and look for the weak points. I'm headed into town."

"Town" was the civic center of Polaris. It was the government nerve center and contained everything in the arsenal he would need for Program Six. He just hoped it wasn't too late.

Two

Colby

Colby Ryder was twenty-eight, single and at the top of her game. It had been like that all of her life. First impressions told people that she was a cheerful, outgoing young woman who flipped her curly tresses at the world. Soft brown eyes belied a scientific nature. She stood tall at five foot seven, slender and straight. Her uniform seemed designed for her. What didn't show was her determination never to dishonor it.

Everything seemed to come easy to Lieutenant Ryder, Astronaut extraordinaire. Few at first saw the competitive streak that fueled her ambitions. They realized soon enough that she possessed a brain that continually sought the best solution, the fastest track, the best burger in town.

Her early years had been spent in Falls Church, Virginia, excelling in a private school for exceptional children. At twelve, Colby showed an aptitude for the sciences of astronomy; also for mathematics and marine biology. She also displayed unusually sharp shooting skills. Her parents were not hunters, but her classmate, Bertie Claxton, and her father loved going to Tennessee to hunt white tail deer. Colby was invited and amazed Bertie and Mr. Claxton with her unerring aim.

"You could live off the land, young lady," said Claxton.

"Shooting comes easy for me," Colby replied.

"Colby makes everything look easy," scowled Bertie. "You should see her in the chemistry lab." Chemistry had been Bertie's biggest problem, but she was grateful for Colby's help. "She's Miss Perfect," added Bertie. It was really mild envy, although Colby had engendered out and out jealousy by other students. She was every teacher's pet and what made it galling was Colby didn't try to instigate this. She was one of those rare individuals who excelled at most everything she tried.

Colby sat at her desk, trying for the hundredth time to reach her father. She had been a resident of Polaris for twenty-two months. The first thirteen months were filled with conversations with her parents back on earth. The vocal communication system had become erratic. Almost everyone on board was affected, and most attributed it to aging equipment at NASA. Her fingers flew across the keys, sending the usual message: PLS CNTCT ME ASAP. She feared that the usually dependable system was damaged and that she might never hear from her parents.

"Tell your dad hi for me." The voice came from behind her.

"He doesn't even know you, Iona."

"Maybe he will." The voice was childlike. Wistful.

"I believe he's involved in more important things right now. I just wish I knew what."

"I bet I could call him."

Colby hit 'send' on her keyboard. "How?" she asked.

The voice persisted. "Did you try the Intergalactic Internet Freeway"?"

"That system's down too, for the past week, Iona."

"Yes, but if you go through the Tools Access and bounce it off the Satellite Underline, you might get lucky."

Colby sat up straight. That was a possibility. She turned in her chair and regarded Iona.

"That's not a half bad idea."

On a small table behind her was a computer about half the size of a shoebox. It was metallic gray and had a round screen. On the screen was a hologram that resembled Colby when she was nine. IONA stood for Individual Optimal Neural Automaton. This high-tech instrument was customized to its owner who also tagged her with "Iona," as if it were a pet. When SysTek Systems developed the first Ionas, they decided the technology was there to give them human bodies, something like the early twenty-first century servant robots, but with more human features. It was even thought that the IONA would be given synthetic hair and eyes that resembled their owners. The developers, a private consortium in NASA's employment, had good luck with the android astronauts they created for deep space travel. The development was given up for practicality. The Ionas were easier to handle as small boxes and could be manufactured at a fraction of the cost. The owner photographs were a nod toward the earlier idea.

"Why don't you try it, Colby?" Iona suggested. "And if it works, tell Commander Addington that I verbalized the idea first."

Colby gave her little computer a level look. "I had no idea you were such a political animal."

"I'm not. I saw it on one of those old movies you like so much."

Colby rolled her eyes and went back to her console. She hit the appropriate keys and waited for a signal. "This might take a while."

"It's not like I'm going anywhere," Iona offered.

Colby had to laugh. "If this works, I'll take you to the fireworks display tomorrow night."

"Oh boy, oh boy. Crash boom fizz! I love that stuff. Can we get cotton candy?!"

"You don't eat cotton candy. You don't eat anything!" Colby said.

"No, but I like looking at it," Iona replied. "It looks…fluff…fluffer."

Despite the advance vocabulary and reasoning microchips that ran Iona, there were always bugs that intruded.

"Fluffy," Colby corrected.

"Yeah, fluffer," Iona repeated.

Colby stared at her screen. She was trying to "will" the message from her father to appear. The anxiety of non-communication with earth had been slowly turning to panic.

She had seen the concern in Jake Addington's eyes. He was cool. The man had earned his stripes in the Intrepid disaster. The Intrepid was a shuttle that was hit by a meteor. Then twenty-three year old Lieutenant Addington was third in command. Even with half the crew dead and the shuttle badly damaged, he was able to pilot it safely back to the Cape. And now, running an enterprise as sophisticated as Polaris had earned him the respect of even more important people in all the fields relevant to the giant operation. Colby hoped that he had a plan for what was happening on the earth hundreds of miles below.

"Anything? I can't see because you're in the way," Iona complained.

Colby wondered who had programmed in the petulance, and then scooted her chair to the side so Iona could see the computer screen. "Better?"

There was a pause. "There is nothing happening," Iona said, her voice downhearted.

"It takes time coming off the satellite." Colby tried to sound more confident than she felt. "What's happening down there, Dad?"

There was silence.

THREE

The Commander

Jake left the Command Center, which held records on all personnel, equipment and storage information. They were going to send Morris some data that would interface with Program Six. Until the report was completed, there wasn't much more he could do. Get in some relaxation, Jake, he told himself. It might be the last free time you get in the next few weeks.

As he strode toward the transit station, his thoughts turned to Colby Ryder. That was a form of relaxation. It was not sexual in nature. He was old enough to be her older brother. At least her father. Twenty-three months ago, just before she got to Polaris, Jake knew all there was to know about the young woman. Her transcripts told about her advanced Agronomic degree from Purdue. From Indiana to the University of Texas at Dallas, she had amassed all sorts of academic awards. Athletically, she was active in volleyball, gymnastics, softball and swimming. There were trophies galore to back up her athletic prowess. She was accepted into the astronaut program at twenty- three. Not only did she break academic records, she set high marks during the physical training as well. When she was assigned to Polaris, there was a great deal of enthusiasm to meet her.

The only thing Jake did not expect was her looks. Oh, he had seen the pictures. They were full color prints of a young lady with inviting brown eyes, an open expression, and long, dark hair that fell to her shoulders. Her skin was lightly tanned, which was set off by faint freckles around her nose. However, when Colby reported to him, her hair was still curly, but shorter. She later admitted that she was uncertain there would be a stylist on board, something she needn't have worried about. She wore little makeup and gave him a direct gaze, almost as if she were at attention. Her voice was confident without sounding boastful.

Jake had a daughter, herself an astronaut, who had been killed a little more than two years before, but he realized that Colby was not a daughter figure. No, he did not feel fatherly the first time he saw this young woman.

Early on, because of his access to her personnel files, he knew something about Colby that most interested men could only find out over time. She was detached when it came to personal relationships. And while she was warm and willing to go out, she exuded an aura that kept one at a distance. It was not aloofness, but rather her drive. She was single-minded and the mission at hand was her focus, not romance. She didn't exhibit a temper, but at times could be a smiling steamroller. Jake knew that the hardest part of his assignment at Polaris would be to hold a tight rein on his emotions. He knew that losing his heart to this comely lass would be disastrous.

Their relationship remained professional, but Jake was able to get Colby to go out and eat with him frequently. There were eleven really good restaurants on Polaris and they both found their favorites. It became a shorthand between them.

"Hey, fish or fowl?" he would ask. That meant The Landing which boasted a large frozen storage assortment of Gulf shrimp and Maine lobster. Or it could mean The Cluck Bucket, which was a popular franchise in the states. They also loved Nino's which offered old Italian recipes and faced Mars every three days. So when Jake asked, "Pasta?" He could count on a willing companion.

Staring out at the horizon from the command center, Jake pondered the problems he faced which were huge, but despite everything he was definitely looking forward to spending time with Colby.

Her call system rang. It was Jake's ring tone. "Hey, Lieutenant." His voice came out from various points of the walls. It sounded as clear as if he were in the room.

"GameTime?" She smiled. GameTime was the amusement center in what was called The Jupiter Quadrant.

"You up for golf or tennis?" she asked.

"What do you think?"

"I think I'm going to wax your butt no matter which we do," she said, and meant it.

"How about doubles?" he asked "I hate getting my butt waxed."

"What time?" she wanted to know.

"Seven? I'll pick you up at transit station 14."

"Dinner afterwards?"

"It's a must," he said.

"Fish or fowl?"

"Fish. Tennis always makes me feel salty."

"See ya at seven," she promised.

Four

Downtime

Colby sat at the transit station, waiting for the bullet to arrive. She put her concerns of worldly matters and her parents aside and thought about Jake Addington. She suspected that her boss was in love with her, or at least had a serious crush. Colby had the utmost respect for the man, and sometimes that respect turned to admiration and sometimes the admiration turned to affection? No. A closeness. He was sexy, she thought, but not her type. What was her type? She hadn't met that type yet. If she did, she missed it.

The train pulled quietly into the station then stopped with a sudden hiss of air brakes. Jake waved at her from one of the compartments.

"Coming, Boss!" she cried with a smile.

Jake stood as she entered the compartment. She waved him down.

"That gentleman's gene of yours really kicks in, doesn't it, Commander?"

Jake plopped back down and patted the seat next to him. "Have a seat, Ryder, that's an order."

"Yes, sir."

"We're headed over to GameTime, did you want to stop for a refreshment first?"

"Nah, I'm refreshed."

"I figured we'd play doubles, did you want to be my partner?"

"Definitely. Last time I got confused and tried to pick up your clone." He gave a soft laugh and watched the interiors of Polaris zip past their window. He was debating on whether he should tell her about what he'd seen on the aero-screen. No. Relaxation. That was the key. He was going to have fun this evening. It might be the last time to have fun for some time.

The Jupiter Quadrant was a universe of neon and bright lights. Jake and Colby got off the train and headed down the blue corridor.

"This one is on me," he said, holding his hand up to the screen. The Polaris economic system did not require cards or money. All spending was done by handprint and the proper accounting was applied to the individual's credit account. The Polaris Enforcement Squad had virtually eliminated stolen identity.

"See ya on the court," said Colby.

He entered the men's tunnel while she entered the women's. The tunnel was dimly lit as Colby undressed. She put her clothes in a tube and hit the button. Then she headed down the tunnel. A low hum started up and she could feel her pores react. It was like being in a huge magnet where it felt like her body was gently being pulled in every direction. It was a pleasant feeling.

"This is nice," she told herself.

Two years earlier, before she came, GameTime had done away with the hologram system. At the time, playing your own hologram had been hi-tech. GameTime had improved their product with temporary cloning. The DNA conductor could clone the individual for twelve hours. In minutes, she would be playing her and Jake's clones out on the court. A rogue enzyme was inserted into the clone so that after about seven hours, the body would start to disintegrate, fading like a wraith. In twelve hours, there would be nothing left.

"I wonder which button Jake hit?" Jake had the choice to make his and her clone three percent smarter or three percent stronger and vice versa. The GameTime people found that this made the competition more fun, because in their early efforts, the clones kept playing their human counterparts evenly. There were few victories for either side. That didn't set well with the competitive crew, which included just about everyone.

In the locker room, Colby chose a white outfit with a ruffled hem. The Wilson racket had a grip that conformed to her hand and she was ready. She checked herself in the mirror and posed.

"They'll eat their hearts out," she said to no one in particular. Funny how accustomed she had become to talking to her robot Iona.

Jake was out on the court juggling three yellow tennis balls. He glanced admiringly at her. "Hey, I think white is your color."

"No sarcasm, please. What do I need to know about our opponents?"

"She will be smarter, but weaker. He will be smarter and stronger."

"Come again?"

Jake shrugged.

"Hey, I like the competition." She shook her head and saw the clones enter the court. The Colby clone waved and gave a friendly smile.

"Hey! I'm looking forward to waxing your butt!" said the clone. The Jake clone smiled too, but had a look of iron in his eyes. Colby nudged Jake.

"She's wearing yellow. I hate yellow."

"Yes, but she's smarter than you."

Now she elbowed him. "Oww!"

"So what do we call them?" she asked

"How about Sage and Thyme?" he offered, digging into his store of culinary phrases.

"Yeah, I like Sage. I look like a Sage. Not the herb," she clarified.

Jake was shaking his head. He pointed to his clone.

"No, he's Sage. She's Thyme."

"You are strange, Jake Addington," she said.

Thyme and Sage won the day after three hard fought matches. The four of them shook hands. Colby was taken by surprise when Sage grabbed her by the shoulders and planted a kiss on her mouth that was a little more than friendly.

"Is that what you were thinking, Commander?" She wiped her mouth without making a show of it.

Jake shrugged innocently. "I have no control over that."

She watched the departing clones and wondered what they did when they went back to the locker room.

"Let's go get fish. You have a hungry woman on your hands."

FIVE

Decisions

The Landing was not crowded as they entered the main restaurant. A wall aquarium surrounded them with all sorts of exotic fish.

"We're in booth thirteen. Superstitious?" he asked.

She knocked on the wooden railing leading down to the dining area. "Thirteen's one of my favorite numbers."

He grinned, took her by the elbow and guided her through the room. He saw a couple of astronauts who had been scheduled for a return on the shuttle the previous week. That mission had been scrubbed and he was afraid they would ask him about it. He steered Colby to the booth.

"Let's get to it. I am famished."

Colby waved her hand over an icon on the top of the table and the menu appeared on the flat surface. She had no idea of what she wanted. "I don't know, do I feel like fresh water or salt water?"

As she scanned the menu, Jake thought about salt water and fresh water. How long would The Landing be able to offer up this food before they ran out? Morris had told him that they had at least a year's supply of food. After that, it would be a crapshoot. They already had Colby's team on Mars, trying to grow natural seed. Success seemed a long way off. "You're awfully quiet all of a sudden," she offered.

He gave her an uneasy smile. "Just glad to be in your company." He saw the look in her eyes. He wasn't fooling her.

She studied the menu offerings, looking away from him. "What is it, Jake?"

"I'm just thinking about things. You know, Polaris and all the details of that. It's a big job. I've got all sorts of..."

"Problems?"

He nodded.

This did not sound like her boss. He rarely showed vulnerability by complaining about problems. This had to be something else. "What are you not telling me?"

He sighed. "Let's eat. It might not be anything." He felt the lie stick in his throat. "I'll take you up to the Command Center. We need to talk."

Colby swallowed the anxiety she felt. Whether she had Catfish or Marlin, it didn't matter now. Jake had some bad news and he wasn't ready to tell her, not yet.

Jake tried to get the conversation back to a normal level. "So what are your plans for Halloween? There's going to be a party in every quadrant."

"I don't know, Jake. What are you doing for Halloween?" The inflection in her voice told him she wasn't going for his attempt at casual conversation.

"I'll be at the Command Center. You have the next week off, right?"

"I do." She offered nothing more.

He continued with his business as usual approach. "Do you think you'll be ready for Mars by April?"

She nodded. "All systems are go."

This sounded good to Jake. They were ahead of schedule. Surface prep had been complete, and the artificial greenhouse in Colony One would be ready in two more months. He saw a ray of hope.

She gave him a curious look. "You're not telling me all of this good news to make my day, are you?"

"No, it looks like things are moving along. We could have natural food in two years, maybe sooner." The words almost slipped out unbidden, but he remained silent. Two years, he thought. It had seemed long, but now, there was not enough time.

The rest of the meal was eaten in relative silence as they both were lost in their own thoughts. They refused dessert and Jake suddenly became the Commander.

"What say we go up to the Command Center."

"Yeah, let's," she agreed. That should change the dynamic.

He tossed his napkin on the table and stood up. "I would like to hear your input on Program Six."

Colby felt her stomach tighten. Among the Astronauts, Program Six was nicknamed "Last Resort."

Within five minutes, they were in the Command Center. Jake strode over to a console and hit some switches. "I guess the best thing is just show you. What you are about to see are past and current satellite shots of earth."

Colby leaned against a table, studying the room. It was not a place she spent much time. When she spoke, there was a worried tone in her voice. "I've been watching earth. And something's going on. I'm not sure what," she added and waited for his comment.

He hesitated. "How did it look to you?"

"It looks like earth. I see it every day," she said, without enthusiasm.

He nodded as the aero-screen came on. "The video is in time-lapse. The pictures go back about twenty months."

"Two months before I got here."

"Right."

The large, round screen showed earth. It was three-dimensional and she felt like she was actually looking at the planet. The picture jiggled a bit to show passage of time. She saw the dates flash forward in the corner of the screen.

"What am I looking for, Jake?"

"Just keep watching."

Many tiny black dots appeared here and there.

"Sun spots?" she asked.

Jake's voice was low and tense. "Keep watching."

What did he want her to see?

The position of the satellite changed to the dark side as it completed its orbit. She was looking at the Eastern Hemisphere. Night. The bright blue marble began to rotate faster. Jake hit another switch and she was seeing things that she'd not noticed before. She was looking at another time, another orbit and the shot showed the Western Hemisphere at night. Her eyes moved to the lights of the big cities. The black dots were a bit bigger.

"Those are the electrical outages we saw," she said.

Jake kept his eyes on the screen.

"Here's what we've seen in the past few months," he said.

The wide, green swath that was the Amazonian rain forest bled into a drab brownness.

"Holy smoke," she muttered. She jumped up and ran up the stairs to one of the many observation decks on Polaris. She looked down on the real earth. How had she missed it? It was not obvious if you watched it with the naked eye, but the speeded up transmission revealed the true picture. "Holy smoke," she repeated. Now she really looked. The blue marble was still blue, but now it seemed to be fading into a drab sameness. She felt Jake's hands on her shoulders. "I never saw the change."

"You look down there and see it everyday. At least you look at it, but you don't see it. I never noticed it either. We've had bigger fish to fry with the blackout at NASA, the Mars program, all of that."

Colby looked stricken. "So you're going to try Program Six?"

He soberly looked at the browning earth below them. "It's our last resort."

SIX

The Unexpected

Colby got back to her room and sat down in her favorite chair, giving herself time to organize her thoughts, and yes, to calm her racing mind. After a minute, she pulled up the files for Program Six. No one had ever expected it to be utilized, especially her. But because it was being called into play, she would be one of the astronauts involved. Her group would shuttle back to earth with necessary provisions to help save the planet. At least that was the plan. Because she had never thought there was a remote chance that Program Six would be needed, she had thought little about it. Right now, her faith in it was weak.

"Colby."

"Not now, Iona."

"But..."

"Not now. Run the outline for Program Six for me." Her mind was on the mission. She was not sure who would be in the first seat. Commander Morris was the most likely candidate. She knew it wouldn't be Jake. Captain Simonton was also a good bet. He had been on Polaris since day one and had trained a lot of Jake's men and women. The TV was playing "Thor," a film about a god come to save earth "I thought that one scared you," she said to Iona without looking at her.

"Colby..."

What was it with her little computer? Sometimes she seemed to have a mind of her own.

"I don't have the time..."

"Your console. Check your console," said the little computer. Colby looked over and saw the code on the screen. 4678KTX. Her father. He had contacted her.

"Your plan worked, Iona!" Colby hurried over to the panels and hit the appropriate keys. All of a sudden, the vocal communications system came on line. She heard her father's voice. It sounded distant, as far away as the small town where he was living. The signal was weak. Going to the tools access and bouncing it off the Satellite Underline had brought back both systems. Maurice Ryder's voice sounded tired.

"Hello, Darling."

"Dad! Thank goodness!"

The voice continued without hesitation. "When you get this message..."

She froze. It was a recording.

"..we will be dead."

Her heart skipped a beat. She went on listening to the message with mounting dread.

"The food supply has been contaminated as you've already learned. The unknown virus went worldwide faster than the AIDS epidemic from the last century. When this thing mutated, the food chain was destroyed in a few months."

Colby found that she was holding her breath. It had been a lot worse than anyone at Polaris had imagined. A nuclear war could not have been as effective at destroying the populace. At first, this menace had come by stealth. No one saw it approaching because no one had planned it.

Once it was identified, the word spread quickly. Not long after she left earth, there were reports of people who were aging faster than normal. In the beginning, it had been tabloid material, showing young, good looking movie stars with age spots and wrinkles on their faces. Colby didn't see any of it first hand. She was already established at Polaris when the mainstream media was inundated with reports about this incredible aging phenomenon. Her father's message continued to play.

"There must be pockets of survivors. I know that there were attempts to store unaffected grain, but to what extent, I do not know. Here, people are dying rapidly. Not only that, but their body chemistry affects them differently. Dementia, insanity, disease of other kinds. The panic has not stopped. Some blame me for this situation, because of my work. It is too late for your mother and me."

Colby did not feel the tear that trickled down her cheek.

Her father's voice seemed even heavier. "This is very important. I have provided a life source for you. Uncontaminated seeds. It's enough natural seed to begin to replenish the earth. I don't want to put you through hell, but I must put you through some very difficult steps. I contrived it that way to eliminate anyone who tries to stop you" His voice disappeared. Colby was on the edge of her seat. His voice choked.

"What? What is it, dad?"

"I'm sorry. This is very difficult for me. I can't trust anyone to find these seeds. You must seek them out and find them for yourself. There are people here who...who are frankly insane. And when you get back here, do not trust anyone. Listen carefully. It will be a life hunt."

Without turning in her seat, Colby spoke to her computer. "Iona, are you getting this?"

"Recording," she replied.

Her father went on, "Father Time has passed his prime and he is slowly dying. Bury him atop the hill where Clover still is lying." Clover. That meant something. But what did it mean?

"Goodbye, my dear girl. We both love you. Your mother can't...she's not here."

The transmission broke. Colby sat like a statue. She was trying to get her mind around the message. Then it hit her. Her father said that her mother was not there.

"She's dead." Colby whispered to herself. "They are dead."

SEVEN

Leading

Program Six was no longer viable for her. Colby had to move much faster than that. She had to get back home to find the seeds. And beyond that, planting them before it was too late, that is, if they would grow. Was the soil healthy? There must be time for tests. Had the virus changed the growing seasons? Would someone appear to help her? So many questions flew into her mind.

She had to talk to Jake as soon as she could. A plan was forming. It was a good thing the military had taught her to think in terms of emergencies. She shook her head. I guess this was the mother of all emergencies, she thought.

First, she would request one of the special emergency modules. There were still individual ones available, she knew, but one option had been removed. Once she left Polaris, there was not enough support on earth to send her back. The truth was, she did not expect to return to Polaris, ever.

She found Jake, still at the Command Center, his eyes focused on current images of the earth. When she hurried into the room, he looked up from his console. He saw her dried tears and what remained, the grief etched on her face.

"Colby? What is it?"

She was determined to be stoic, something that had gotten her through many a situation through the years. When she spoke, her voice sounded as if she were offering a report. "I heard from my folks."

"What did they say?"

"My father. It was a recording. They're both dead."

They talked deep into the night. Jake did not want her to go alone in the module. He knew he could order her to take the shuttle with Program Six, but he also knew that sending her in alone was what she wanted. He was confident she would be as effective on her own and at some point she could reconnect with the others who would be a month or more behind her. Still, the idea of her alone down there, with no help, did not please him. He attempted to dissuade her.

"Remember, your father isn't the only one with good seed."

"I know, but..."

"The government has silos in Kansas. It's in a restricted area and we have no reason to believe it's been compromised."

"So that's what the team will go after?"

"That one and others."

She couldn't believe that Jake was handling her. He should know that she was trustworthy, but he didn't seem to put his faith in her. She knew that there might be good seed in other places, but the stash in Central Texas would be the most accessible. "Jake. I've got top clearance. What is it you are not telling me?"

He rubbed his eyes and blinked up at the ceiling.

"Ah, hell. I guess it doesn't matter anymore. We're playing an entirely different game now." Jake took a breath before speaking. "There is a cave in Norway. It is man-made and it contains uncontaminated seeds of all kinds. Supposedly it's still safe."

"Why supposedly? It either is or it isn't!"

"Before we lost contact with the President, there were rumors that some of the stronger victims of the virus might have organized themselves enough to breach the Norway site and maybe the other few remaining spots. They would've been the younger population, but that's not true. They didn't have the strength to go that far and most of them died. Remains to be seen what the status is now. There are always terrorist threats even here at Polaris. They've gotten through. But this is different."

What was he withholding. His eyes stared away from her, out into space.

Colby thought back to earlier dispatches. "Victims organizing. Doesn't sound likely. But maybe terrorists? Still, they'd run out of strength as well." Nothing made sense.

He nodded. "You've got to respect the victim's convictions. If they thought there was a cure, why not go after it? And there were some who thought God wanted to destroy the earth, allow the corrupt seeds to create havoc and destruction, even though the disaster clearly was man-made. But no one is spared. Everyone ate earth's foods and apparently grew old at an alarming rate. I can hardly fathom it."

"It would be hard to get to Norway now. Too many of the systems are knocked out, air travel, ship." She hesitated, but not for long. "So how about it, Jake? Will you grant my request to go ASAP?"

"Like I said, we're operating under a different set of rules now. Do you realize, there is no United States government operating at present or at best, a remnant of it. The Senators and the President were among the first to die. Our mission is to restore a government, if we defeat the virus. Will we choose it to be the same? I imagine so. I pray you can solve the riddle. Your orders will be ready within the hour. I'll sign them."

She saw a shadow cross his face.

"I also realize I could be signing your death warrant."

Colby laughed. "You also know if I don't get back home, we'll all be under death warrants."

He allowed a bitter smile. "True, true," he murmured softly. Then he was back to business. "So what can I give you to make your mission successful?"

"I need at least six months rations. When I get down there, I can't eat anything that comes out of the ground."

"I can give you six months worth, maybe more."

She shook her head. "If I run out in six months, it's all over anyway."

"Then I'll give you a year's worth."

"I don't want anybody shorted."

He was already shaking his head. "No one is going hungry. At least not for another couple of years up here. I'm also sending the team to implement Program Six. If they can get down there with good seed, plus whatever your father has hidden, we've got a fighting chance."

Before she could say anything else, he reached over and kissed her forehead. They held an embrace.

"I'll always be in your debt, Jake," she whispered.

"Right back at you, kid."

Kid. For a brief moment, she felt like a child. For an instant, she almost took back what she wanted. Maybe it would be better to go down to earth with the team. She didn't fear going home alone, but was it the right strategy? Would it have been wiser simply to turn their backs on earth, declare it a disaster area and never return? Only God knew the answer to that one.

EIGHT

Earthward

That night, Colby stared at the lightweight gray material around her bed and thought of the enormous Polaris space station enveloping her, protecting her and her co-astronauts from the hostile world of space outside. Light emerged from the hidden fixtures set into the wall. Minute particles seemed to float in the light, as if swimming in a benevolent stream. There was a quiet unlike anything on the earth, the sophisticated machinery powered by acoustically insulated on-board generators fueled by the sun. They would last as long as there was a sun. If there was sound to be heard, it was of music, conversation or the cartoon noises of computer games in use throughout the craft. Even the bullet trains ran silently. The power station emitted vibrations that fanned out and re-routed meteor showers from itself and earth.

She could not live without the incredible platform that was itself cutting through the dark cold space like a laser in the blackness. Colby was aware that Polaris was visible from the earth, and that it had become a sign of hope, as the North Star had always been for navigators, that there was some remote chance that everyone was not doomed. To those on earth, the ship's light appeared as if it were a star, but moving inexorably through the sky, soon disappearing from sight in one sphere and appearing in the next. Once Polaris had been poised to cause terrible destruction as a Navy missile during an uneasy period called the "Cold

War," but that program had ended, and the astronauts had quickly co-opted the name.

For a short period of time, the astronauts had enjoyed their status as cheerleaders for the earthbound, and had sent them daily messages of hope. Now, the truth was a harsh reality. Few were left on the earth to see the man-made star. Instead, compelled by strange and unexpected circumstances to change their focus, the brave explorers aboard the sophisticated craft turned their faces toward Mars, an inviting beacon. The Red Planet was already peopled, although they were few in number. It had become the launching place for expanded missions out into the universe with great plans for the future.

Colby had fully expected to see the Red Planet personally, to touch the surface of the newly explored neighbor to her own earth, to stay for two or maybe three years in the elaborate new outpost. Now all that had changed. She had learned this morning that there were fellow officers aboard who would fulfill that mission, but it would not be her.

Colby's long-range strategy didn't matter now. Instead of following the path she had planned, trained extensively for, she was scheduled on the next shuttle home to be thrust into the middle of the terrible troubles on earth. She frowned. This must be how sacrificial lambs must feel, she thought. Added to that notion was her own belief that going back was a futile effort. It would gain her nothing, nor would it help those on the doomed planet that was her home. What would it be, she pondered, but a burial detail? And after that, who would bury her, she wondered.

She thought of her father. Had anyone tended to his burial? Surely someone had. He was Dr. Maurice Ryder, a scientist at the highest level of government, an international agronomy expert armed with two PhDs and literally hundreds of grants worth millions of dollars through the years.

She did not question that she would go to him as quickly as she could, her goal to solve the mystery of the grain. Once she received the command, she was ready to go. Even as the others would soon be hurtling through space toward Mars, the dream that had been her greatest hope, to be with those on the great journey, was about to be dashed. For once Colby became earthbound, there was no way she could rejoin the voyagers.

Too much chance of contamination from whatever was decimating the earth, and even if that was resolved, she'd be at the back of the line for the new flights, already full of eager young astronauts, with technical degrees from the world's finest universities, strong, virile officers in the prime of life. Not only were they strong enough to make the voyage, but at the right time, though unspoken they could also repopulate the planet Earth.

She would miss the excitement of the unknown, of the possibilities, of going beyond what previously had been experienced by mankind. Above all she would miss her mentor and friend, and yes, if she had stayed longer, perhaps something more, Colonel Jake Addington, Commander of the Polaris Space Station. Her unspoken fear, even beyond losing her hard earned position in the astronaut corps, was that she would never see him again.

Still, her father's wishes could not be ignored. He was a scientist first and foremost, and he always had a reason for his actions. She had always trusted that. In the same vein, she had learned to follow orders, which was what she was doing to the letter.

Colby sat in the soft light for a while, and then flipped on a tiny television and stared at it mesmerized. On the screen, she studied a replayed transmission of the visuals of the earth that Jake had shown her. As she watched, a wide green swath of the rain forest bled into a drab brownness as if a painter had covered the canvas. The image before her was familiar, and as she somehow knew, nothing had changed for the good. She watched a moment longer, and then flipped off the monitor. Shortly she would be reentering the now alien world. There would be plenty of exposure then. She wondered if there were enough technicians left to land the shuttle, but surely Dr. Ryder had considered such a thing. He wouldn't summon her to an instant death.

The silent gray screen reflected Colby's own image back at her. Slowly she walked across the faux wood floor and finished packing her sparse belongings. Despite her deep disappointment, she was resigned.

Somehow, she had fully expected to be called home. It was just a matter of when it would happen. Now she wished that it never had.

Funny how in her youth the seasons would flow past, spring and its promise, green and alive with winds and storms and bright sunny days; summer, slower in Texas, hotter than she liked in July and August, but aware that it would be over and autumn was coming, sweet relief from the blistering sun, with cooler days, football games and college. But now it did not matter what the date was. She would go into the winter season, for despite the familiar categories of time and place, everything was out of kilter.

One thing she liked about the city-sized Polaris space station was that it was run on a schedule, everything in its place, whether time, belongings, departures or arrivals. Because of such order, she would be on her way at the exact time allotted. No chance for someone forgetting to call her or herself forgetting. None of that.

She considered how strange it was that the military loved order at every level, and yet at the beginning of a skirmish or full blown battle, when the first shot was fired, order went out the window and chaos ensued, except for those churning out orders that were meant to overcome the chaos. Most military people would rightly say the learning of order, of a chain of command, enabled leaders in a battle to think clearly and bring the situation under control. That was the military theory at least and the work that NASA did owed everything to that belief system.

Before she prepared for her journey, she and Jake had worked out a plan of communication. The military had taught them both to use secrecy and common sense. Whatever was going on down there, Jake needed to know, even if it was terrible news.

Jake had waited with her. It was like a scene from an old war movie, when a couple has only hours left. The pair went to the Polaris Club and had dinner and coffee. When the music started, Jake grinned and invited her to dance.

The resident DJ turned on his antiquated hologram machine, and the pair was dancing with their look-alikes, which made them laugh, until the time for Colby's departure.

At the last minute, had put his arms around her and held her in silence. She stared up at his intelligent face framed by salt and pepper hair, his inquiring gray eyes, and the serious expression on his face. Already on the shuttle were her gear, her computer she called "Iona," a nickname that offered the suggestion of friendship, and finally her solar powered four-wheeler.

Colby was amazed at how much useful equipment could be packed on her compact vehicle, everything from a tent, to a taser, to a year's supply of uncontaminated special food and water distilling equipment, tools of every kind, and of course the intergalactic global positioning device where she could be located should communication be lost.

There were binoculars and a lightweight insulated tent developed by top mountain climbers, complete with an attached sleeping bag and mattress. She removed Iona from her packing and strapped her into a special damage-resistant backpack.

Her eventual destination was classified top secret, known to only a few, although the secrecy seemed almost purposeless this late in the day, like battle strategies after a fight.

She was heading for the Houston Space Center, an amazing collection of structures rebuilt after the worst hurricane in history had almost leveled it a mere seven years ago. Now, storms couldn't touch the new edifices, but the insidious virus had left only a few people that she was aware of to operate the extensive scientific equipment. She knew she could bring the shuttle home alone, but it was easier with the controllers helping her. She hoped they were still there.

Her journey would finally take her to her parents' home base, if anything remained of the small farm community of Rice, Texas, in the heart of the state. It was there that the United Nations had joined forces to build an agriculture research facility, of which her father was the superior officer.

Jake refused to leave her right up to the last hour before launch. She boarded per instructions, but even as the hatches were being prepared to close, before she stepped inside, he slowed the workers down long enough to give her one more item, hidden in a briefcase. It was a book which looked old, but it was a 1993 reproduction of Thomas Jefferson's

"Principles of Gardening" book. At the front was a note from the brilliant President and farmer: "The greatest service which can be rendered any country is to add an [sic] useful plant to its culture."

She thanked Jake for that as well, and finally boarded the shuttle. He stepped away and walked back to the viewing area. The expression on his face was sad, but beyond that she could not understand what was going on inside his mind.

She didn't know what she would find when she got there, or worse, what would find her. She was certain of one thing, that it would not be a round trip journey. Within the hour, she was hurtling earthward.

NINE

Sam

Dr. Sam Michelson had come a great distance. When the enormity of the disaster became clear to his crew stationed at the South Pole, decisions had to be made. He could have stayed there with several hundred others and hope that the virus played itself out, but he made a different decision. He wanted to go home. His superiors said yes to his request. There was little need in retaining him. Besides, at some point supplies would run low and finally disappear.

It had been devastating to see the international laboratories carefully built deep under the ice slowly grind to a halt, the brilliant scientists who had answered the question of the substance of Dark Matter suddenly with nothing to do because their governments were dying and gave them no further instructions. Some had gone back to their homes. One or two ended their own lives. Some simply waited.

Sam caught a ride on a cargo plane that had delivered a year's worth of supplies and headed for Santiago, Chile.

What they found there was devastating. The bodies had piled up quickly and there had been too few left alive to bury them properly. Many were lined up at the airport, decently covered with blankets, some on camp cots, but some simply sitting in the waiting areas, slumped over in the large comfortable chairs there. Sam guessed that he would find the same scenario wherever he went. But hopefully there were some alive, some immune to the scourge.

His ride left him there, flying on to an isolated base in the Australian Outback, hoping against hope that some survivors somewhere had the answers. He found an empty recreational vehicle and spent a relatively quiet night in it. Early morning found him walking alone on the tarmac, looking for an airplane that he could fly home. He found a small jet that was properly fueled, and decided on that. He hardly noticed the presidential signature outside the cockpit, and the flag with its single white star, red, white and blue panels that identified it as belonging to the Chilean government. The important thing was he was familiar with this model aircraft and had flown one like it several times.

None of his lofty institutional ideas mattered now. There was nothing left to prove, not that he could create a genetically superior corn or that a grain of wheat might be modified to produce ten times what it had always done. In place of all that was the desire to head for home.

The plane felt good in his hands. He flew from Santiago to Quito, Ecuador, refueled and went on until at last he reached Mexico City. He refueled again, and eventually landed at the Naval Air Base in Fort Worth, Texas. He left the jet sitting on the runway, among a score of others that had been abandoned.

He quickly found a cycle, and nearby new battery units. There had been fuel and shelter along the way, but no food. He had figured that would be the case and he had enough dried packaged provisions from the Pole to get him to his destination. He was sick of the dried food hydrated with clean water, its flat taste and brown appearance far from fresh, which he figured he would never taste again. He cleaned the windshield and glanced around yet again for some sign of life. Before, he had been hopeful, but now he expected nothing at all, not a human, not an animal, even a cat to be in his path. He was not let down as he had been for the past five days.

He pulled on his black helmet, snapped the strap tight and headed south on the interstate.

What he had experienced shocked him. Ironically, Sam had always wanted to see the Galapagos Islands and the life forms made famous by the drawings and writings of the 18th century scientist Charles Darwin, who in Sam's view certainly must have been a patient man. As a crop specialist for the United States Department of Agriculture, Sam had waited for plant growth on many occasions, depending on nature to help him with his doctoral thesis describing how certain plants adapted to their environment.

But even Darwin must not have witnessed change in overdrive. It was if people were driving a car with airplane horsepower and could not control it. The result was hardly a welcome sight.

Riding through the evening, alone on the six lane highway, he knew there was no going back. A return to the South Pole was impossible, and there was no place that he had stopped in between that he cared to remain. Home was as good a place as any to…to die, he thought and then dismissed the idea. No use dwelling on it. He didn't expect to have the power to change anything.

All that was left for him now was to travel the final miles to home. If he were lucky, maybe someone would be left, for a little while anyway. Perhaps he could find some kind of food as well. The cycle pulsed to life and Sam Michaelson pulled away from the silent station.

The tall angular young man, his face set in sadness, rode on. There was nothing he could tell himself that would help. His once raucous sense of humor had disappeared thousands of miles and dozens of months ago, left like trash behind him. What good was humor if there was no one to enjoy it. In the days past, Sam had fixated on the devastation he encountered. What particle of hope he had retained was gone now. There was nothing he could do to fix it. He set his eyes on the horizon and made for home.

TEN

Earth Under Attack

The module's new design was a streamlined teardrop far superior to the old chemistry class beaker appearance of the early Mercury capsules NASA had developed. It had proved a dependable mode of transportation for a number of years. It could land light as a feather, hands free. Colby looked out her window and saw the approaching earth. It wouldn't be long now.

"Are we there yet?" asked Iona. The little computer was plugged into the navigation system. Colby gave a light laugh.

"Cute."

"What's cute?"

"You, asking the eternal question human kids have asked through the ages."

"Oh," the tiny voice piped up again. "So...are we there yet?"

Colby felt a chill. Iona shouldn't be asking that question or any other about the landing. "Where do you think we are?" The chill got colder when Iona answered her.

"Latitude 19 degrees, 43 minutes north; longitude one hundred and fifty-five degrees, five minutes west." Colby accessed her monitor and rapidly entered figures on her key pad.

With hands steady as a neurosurgeon, Colby unplugged Iona from the navigational system. "Let's take a look at you." She took the small screwdriver and opened Iona up. The little computer coughed.

"System breach! System breach!"

"Be quiet. Let me see." Gingerly she lifted the outer titanium board that protected the robot's nerve center and peered at the tiny encasement that contained the microchip. There had to be something loose. Then she saw it. A wire had been exposed. With her heart pounding, Colby took a tiny tool from her kit. The tweezer-like device also had a microchip on the tong parts. It searched for a nanosecond and quickly reprogrammed Iona. At least Colby hoped the effort was successful. After a long, anxious minute, she put the tool away and screwed the backing on and plugged Iona back into the navigation system.

"Iona? Are you okay?"

The computer did not remember her confusion.

"I feel fine, why?"

Colby was taking no chances. "Could you give me those re-entry numbers please?" The numbers appeared on her console. They were different from the ones Iona gave her earlier. These were the right numbers.

In moments, they were pulled into the atmosphere. The aircraft shuddered as it defied the pull of gravity and the searing flames of re-entry which seemed to go on and on, but then the sense of clear air around the craft, smooth flying. They touched down on the long runway. There was a moment of relief as Colby leaned forward in her seat as the module glided to a stop on the tarmac. The doors opened automatically, and she started down the steps. The Houston Space Center looked as if it were abandoned, except for a single figure sitting in a golf cart. She walked toward him. He was leaning against the steering wheel. He was almost dead.

ELEVEN

Last Words

A cold north wind propelled dust across a winter brown vista of dark fields, fragments of dry yellow corn husks adding pieces of light, but under that, acres of unharvested corn lay rotting in the cold November air. A grain elevator caught the rider's eye, the tall metal tubing with black skeletal arms against the horizon, dominating the landscape like a monster in some mid-1950s Japanese film. The Polaris crew had loved watching those, laughing and eating popcorn while drinking "near beer."

The four-wheeler cut through the icy wind at more than seventy miles per hour on the two-lane blacktop, its single rider hunched low behind a bug-spattered Lexan windshield.

Colby stopped the vehicle and looked around. Slowly, she removed her helmet and goggles, and pulled her insulated jacket tighter against the unrelenting wind. The grim expression now outlined in Texas dirt belied her beautiful face. Her intelligent eyes surveyed the silent empty Blackland Prairie like a young deer, moving from side to side as she carefully drank from a lined canvas canteen.

At last she picked up a miniature machine much like the small phones of the early twenty-first century from the vehicle and walked to the massive sheltering trunk of a large live oak tree. She fished in her pocket for a tiny titanium disk and slipped it into the machine.

"Go ahead, Iona," she said.

"All right, Colby. Here are the case studies just as I recorded them." Colby stared at the screen as Iona, her customized miniature computer, began its transmission.

There was a buzzing and sparks of light as the familiar message was re-laid. At first Colby wasn't sure what she was seeing. A well coiffed, glamorous woman appeared to be staring at her own face in a television monitor, touching the screen before her and with her finger tracing the reflection of deep lines in her face, and then touching her real face and finding the same etchings. The alarmed woman quickly turned off the disturbing image. Immediately a different picture appeared.

Fans in full throat bellowed endlessly at a football game. The camera found the familiar quarterback as he back pedaled for a pass, tried to throw and instead fell on the bright green synthetic turf, although no one had tackled him. He could not even get to his feet. His teammates hurried to help him. A camera zoomed in as he held his hands up before him, and they began to twist as if with arthritis. He frowned in stunned, disbelief.

Colby glanced away from the screen and then at her own gloved hands. She pulled off the gloves. Her hands were slender and youthful. At the back of her mind, she registered that the turf was artificial, else it would be brown as all other grass on the planet.

She looked at the television again and saw a small child running and falling down and then running and falling again, and then the child began to cry in frustration. At closer examination of the projection, the child had a wizened look. The final image was a man working at a computer who laid his head down on his desk in weariness. His hair turned white and the screen turned into static as Colby watched. There was more. She waited.

Iona spoke now over the static in the voice of a man. "Listen!" it said. That brought Colby to full attention.

"Dad?" Colby whispered as she listened to a repeat of the message she heard while she was still in Polaris. There was more information now. Somehow she had expected her friend Jake.

"Goodbye, dear Colby. The food supply has been contaminated as you've already learned. First, the grain was destroyed by an unknown virus. They only meant to enhance the food supply, but the virus mutated and quickly destroyed the food chain. There must be pockets of survivors. For your sake I pray so. Here, people are dying rapidly. They blame me and the government I represent, their own government. It is too late for your mother and me, but I have provided a life source for you. Listen carefully. Remember, Colby. Trust no one!"

The tears began to pour down her face leaving shallow furrows of damp sand. The weeping was as much from weariness as from hearing her father's voice. Is this what it means to be human, she wondered, and remembered her past ability to hold her feelings at a distance, as if her heart had an icy splinter at its center, a way of seeing the world that enabled her to push on, no matter what. Perhaps after all she had gone through, all that she had seen coming home, the sliver of ice had at last melted. She leaned back against the ancient tree as if she might never get up. It would be so easy to stay there, to let herself join all the others who had gone before her.

The robot's voice pierced the wind. "You need rest, Colby. You need to sleep. It is three days now since..."

Colby wiped away the tears. She was tired, yes, and at one point, she remembered the necessity of keeping up her physical regimen on the Space Station, and she had done that. There was no sign of atrophy in her muscles. Nevertheless, she felt a deep exhaustion, more from the reality of what she had found than her physical condition.

She demanded of herself that the coldness of her heart remain. She had always prided herself on her objectivity. She was a 'hard-ass" according to her friends and even her professors. She wanted no one to see any emotions from her. No tears. No frowns. She got to her feet and started back to the four-wheeler. "Couple more hours and we'll be home. Then I'll sleep."

"Do you want to check your face in my image reflector?" Iona asked.

"Image reflector?"

"Yes, to primp, put on your 'ipstick.'"

"Oh, the mirror," Colby finally understood. "And it's lipstick."

"I see. Do you want to see yourself?"

"No, thanks. We need to go on."

"I have never visited your home before, Colby."

"That's because I've only had you three years and I haven't been home in five."

With that Colby mounted her ride, cranked the engine and rolled silently onto the highway. She still had a good way to go. It was still hard to believe that nothing moved. It was as if she were in a dream without end, in which she alone inhabited the earth. Who had not dreamed of the same horror, some Armageddon, some end times, where everything that mattered was lost. Except this time the dreamer had wakened and it was worse than any nightmare.

She fought to keep alert. She set the little robot to play music, loud, raucous songs she had never listened to riding through the cold beauty of space. Now, it was a connection to life for her.

Something caught her eye then, and she turned her cycle into a lane leading to a faded red barn. On its caved-in roof, a sign painted in black read: "SEED FOR SALE."

TWELVE

Ghost Voices

Colby left her cycle and approached the barn. Nothing stirred, even a barn rat that she might have expected in the old building. There were hay, farm tools, a new looking John Deere tractor that had a layer of dust covering the exterior of its air conditioned cab, and the skeletal remains of a half dozen cows. Outside, she looked toward the house and approached it. She stood on the front porch near a swing with rusty chains holding it in place and creating a rasping sound in the wind. The house appeared empty as well and she had no desire to go inside. Something unseen seemed to be waiting, a pulse, the spirit of the doomed family who had lived there. She shivered in the growing chill and got back on her ATV.

She continued along the highway, and soon spotted a yellow school bus parked at a stop sign as if waiting to go forward. For a moment she let her guard lapse. If school buses were still rolling, there must be life, she reasoned. She moved closer and stared inside.

Dust and grime clouded the windows and she knew without seeing that it was vacant like everything else she had observed. She circled the bus and then parked near the open door. She stepped inside, and realized she would have to blow away the thick particles of dust and knock down a spider's web with the spider hanging dead in the web. Abandoned backpacks and lunch boxes littered the center aisle, as if the occupants had left in a hurry. What a nightmare, she thought. Even the children had to suffer. Had the parents outlived them, or had it been the other way?

Colby reached down and picked up a child's bright pink hair ribbon. The ghostly laughter of children echoed from somewhere. It was as if the small riders were still in their seats watching her, counting on her to stop the scourge. They frightened her, more because she was too late to help them than fear, for they were as powerless now as they had been in the face of the virus.

She clutched the ribbon, quickly folded it and zipped open a pocket where she carefully placed it. But then she backed away as swiftly as she could from whatever was there, the children's presence, their ghosts, the horror that they experienced. The idea of all that sent her clambering backward and she almost fell down the steps.

Once on the ground she realized she had been quite calmly contemplating that every child in the county must be dead, and the detritus in this bus all that was left of them. It can't be, she thought. Such a terror could not happen. God would not let it happen. She realized she hadn't thought of God's involvement in a long time, but somehow the reference came to her unbidden. All those years in chapel, she had been listening, perhaps absently, but still she had absorbed the Word. Wasn't it too late for Him now? Hadn't he left us to our own devices, and we self-destructed, she thought.

She hurried back to her cycle as if someone were walking fast behind her. She struggled to start the machine, missing the keys, jabbing at them as if her life depended on getting the combination right. But then she stopped her effort and climbed off, disgusted with herself for her fear. If only there was something or someone to fear, but she was totally abjectly alone. Wasn't that worse!

She looked up. Even the birds were gone.

Iona piped up, "I want to explore with you!"

"No, you're staying with the cycle. So if anyone approaches you can sound the alarm."

"There is no one."

"You'll do what I say!"

Thirteen

Approach

The November wind was deafening, louder than the roaring winds of the South Pole during storm seasons. It pushed hard against Sam and he pushed back with all his strength, driving the machine as hard as he could into the teeth of the shrieking gale.

At times he would grow weary struggling with the incessant thrust, and he would find a place to escape the awesome raging storm. Had what happened changed even the seasons? Lost vegetation that once might've slowed the wind lay flat and dead, and the wind had won? Shelter was an empty hotel room, an abandoned house, the inside of a crumbling building in downtown Waxahachie, Texas, anywhere that would give him a moment away from the howling living monster that had no beginning and seemed to have no end.

Finally, in Ennis, north of Rice, in central Texas, he had turned down Main Street, the town's central thoroughfare. It was familiar. Hadn't he ridden his horse down it in the community college's Homecoming parade. In those years, the small city had been the limit of his universe. If he had known then how far he would travel, he might've paid more attention to everything around him.

Still he could not remember a wind such as the one he was experiencing. It had blown down signs, taken the roof off a building, blown over tall lights at the football field. But as suddenly as it had started, the living creature's anger peaked and stopped, as if someone had thrown a switch. Instead of the roaring wind, there was a blessed silence.

Sam was grateful for that. He was nearly home.

Outside of the city limits, Sam saw something moving, something unexpected, or was it an illusion? He slowed his bike, set the engine on electric mode which was silent with little vibration to let him know the machine was running. But it was and, he rode in a circle, keeping his distance. He didn't want to frighten the other rider. Was it one of those mirages people dying of thirst in the desert saw? He rode closer. No, it appeared to be a real person.

To his surprise, it was a woman standing near a giant oak tree working with something on her vehicle. She had not seen him yet. His heart skipped a beat. It was someone alive, he thought, something he had not allowed himself to consider. Slowly, he headed in that direction.

Beneath the tree at that moment, a terrible siren set up. Colby grabbed her ears. "Not now. "

"But Colby, look to the east," said Iona.

———◆———

Colby shaded her eyes. The unexpected sight of a cycle zigzagging toward her filled her with fear like nothing she had ever felt. Her body and mind were frozen where she stood. She did not have the strength to defend herself, even if she had wanted to.

Colby had gone through the most rigorous training available to become an astronaut. She had learned to pilot different kinds of aircraft, and been good at it. She had jumped from airplanes, run through obstacle courses, taken all kinds of verbal abuse from her drill instructors, graduated from college magna cum laude and been accepted for astronaut training. After all that, she finally blasted into space, leaving a great many envious classmates behind. It had not occurred to her to be afraid, until this moment.

She held her breath, watching the rider near. For one of the few times in her life she was uncertain what she should do.

The rider drew closer and circled slowly as if he were stalking prey. He had the advantage since he had spotted her before she saw him. Nothing to do but wait and see. Where was the weapon guaranteed not to kill? But before she could think where it was, let alone grab it, he pulled up a few yards from her and planted his feet on the ground. Slowly he removed his helmet and the mask that had kept the flying dirt out of his dark eyes. His face, like hers, was ringed with grime. He must have come from a long distance.

A handsome athletic man in his late 20s stared at her without speaking. What he did next startled her. He grinned at her, somehow too casually given the gravity of the situation, but the searching eyes gave him away. They were not smiling.

She tried to appraise him as he approached her, but she had no resources left. Iona was right. She was bone tired.

He extended his hand. "Sam Mickelson. Sam."

Colby did not accept the hand shake, but took a step back. "Colby Ryder."

"I'm hoping you have water. Ran out yesterday."

Colby kept her eyes trained on him, but unhooked the water bottle at her belt and tossed it to him before he could come nearer.

Sam caught the flask easily. "I trust it's safe."

She nodded in the affirmative although it was hardly audible. "Uh huh."

He drank deeply. He smiled, and started toward her with the canteen.

Colby tried to read him, but she could not. She stood there, swaying in her weariness, and then she began to back away. Something was telling her to escape him. With that, she turned and started to run, toward the bus, the swings, anywhere away from this man whose purpose she did not know.

The stranger ran after her and caught her easily. She kicked hard at him, but he pressed his hand against her neck and she fell unconscious. He caught her, and then he swung her into his arms and started toward her vehicle. His cavalier expression was gone, replaced with something else, a deep sadness.

"But you? Who are you?" he asked.

Fourteen

The Old Ones

Less than six months before Colby and Sam began their separate journeys to the community of Rice, a man named Angel Tikal had begun to realize that something was terribly wrong with his quiet and orderly world.

By the time he knew what was happening, all but a few of the farm houses around Rice were dark and had been for two or more months. Many of the former residents, his neighbors at the cotton gin, the stores downtown, the Parent-Teacher Association, had died of the virus that was lethal in months, sometimes in weeks.

Harry Rose, who owned the grocery store and had personally and unknowingly distributed tainted food, had died, as had his wife and his wife's brother, Kenny. Their cousin attempted to run the store, but he had soon died as well. Those left had cleaned out the stock, most unaware that they were stealing poison.

A few of the town's people had lasted longer, time enough to bury the dead. They used the machinery at hand, ditch diggers, hydraulic lifts, a small bulldozer, and at the end shovels to finish the task. No new stones were set. Angel and the others simply put down bricks or flagstones and wrote on them the person's name.

There are always sturdy souls in a dying society who have the strength to live on within the lost community. Throughout history, because of an innate immunity to the current plague, whether it be Ebola or bubonic plague or cholera, and even if they found no joy or hope in remaining, still they lived on. So it appeared to be the case with Angel Tikal who had dwelt in the tight knit community since he had been born in a sturdy outbuilding on what later would become his family's own small farm.

Angel was the son of an undocumented, most called him illegal, immigrant farm worker from Guatemala, whose last name became Tikal, after the city of his birth. His parents spoke broken English, but they knew that if their child was born in Texas, or anywhere in the United States, he would be a U.S. citizen. It was a law that no other nation had passed, the idea behind it to guarantee that slaves brought to the country in the 1700s and finally freed in the 1860s would have full citizenship. But like many parts of the Constitution of the United States which were written for one reason but evolved to another, the law became a way for strong determined souls to make their way to the "new world" and add their strength to a land built by hard work. And so Angel was born on a July evening, a son of his new country. The joy would be short-lived.

Over the years, Angel's father, Jorge, worked hard and made a name for himself as a handyman who could fix anything. The small, soft-spoken man taught his son in the same way. They were useful to the community, and at some point purchased a fifteen-acre farm that had an old farmhouse on it. They moved from the crude outbuilding into the farm house, itself old, but clean and warm.

For more than a decade they worked in the community and on their own property, building a pleasant homestead. They sold goat's milk and goat cheese. They had a booth at the summer Farmers' Market where they offered fresh green beans, tomatoes, watermelon, cantaloupe, new potatoes and okra, all grown from seeds the family brought from Tikal and every year, used anew. In the fall they would butcher a hog, and that meat would last them through most of the winter. The family worshipped at the Catholic Church that had grown steadily over the past decade, and each child was baptized there.

In school, the children were quiet, as they had been taught, and Angel turned out to have a gift for math. He had been encouraged to attend the community college, but had turned down the opportunity because his family needed him to work. He had dropped out of high school at sixteen, and from that time had worked with his father. His goal was to become the best handyman in the county. He managed to enroll at Navarro Community College to study air conditioning repair and welding, where he excelled in his classes.

He had married a young woman, only sixteen, whose family had moved down from Dallas to escape the violence there. They had two children which the parents raised carefully. Within a decade they fit into the community of Rice as if his family had always lived there. He had his own high hopes for his children to be educated even though they were barely beyond being toddlers. And then the virus came.

Angel was twenty-two when it started, although he had not paid attention to the news bulletins and the fears voiced by others in Rice.

It was on one of their jobs that Angel first noticed something different about Jorge.

Almost overnight, his father's silky black hair began turning white, and it had lost its luster. Angel commented on it, but Jorge brushed off the notice, saying that he was growing old, that was all. Jorge was just fifty.

Angel reminded him of that and teased him a little. But he thought that perhaps his father was ill, and he urged him to have a checkup with a local doctor, which Jorge refused. He had never spent money for doctors, and he didn't expect to start now.

"All the same," his father said, "All the same I am growing old. Look at my hand," he said, and held out his hand which was holding a hammer. "Look," he urged.

Jorge's hand was shaking like that of a very old man.

Angel stared at his father's hand, and then at his own, which was strong and steady. He did notice that the dark hairs on his arm and hand were lighter. From the sun, he reasoned. From hours in the sun, he insisted. But Angel said to him, "You are one of the Old Ones now."

A brief two months later, Angel sat on the front steps of his porch in his regular place. The flowerbeds on each side of him seemed to have settled into the brown earth, and nothing had been planted to replace them. That was not unexpected. His mother had tended the garden out back, and the flowers at the front of the house, and he had just returned from burying her. Like his father who had died months ago, his mother turned into an old woman almost overnight. She had died of congestive heart failure two days earlier. The thing was that by now, everyone knew what caused the aging and its resultant destruction of the human race. The horrible virus that was sweeping the earth seemed to go on and on. Someone had to have an answer to it.

Angel got up from the stoop and turned to the house. He must find out how many survivors were in the county, and they must confront the professor and his educated wife who lived in town. They must have brought the virus in some of the plant experiments they were working with in that huge greenhouse they'd built when they first came. Who else could have done such a thing, although it was unclear why anyone would.

Within a few minutes, Angel had changed out of his Sunday clothes, put on his jeans and work boots, and walked down the driveway toward the street. Slung over his shoulder was his loaded deer rifle. He moved like an old, old man.

FIFTEEN

Sam Helps Out

He whistled softly and picked up firewood and fed it one piece at a time into the fire. It had occurred to Sam that he had smelled no other wood smoke, when in the past the late autumn air would have been full of it, inviting and sweet, oak and mesquite, symbolic of the season.

Now and then he glanced toward the silent tent, a slightly worried expression on his face. But then he would walk a little way from the camp and pick up more wood. At one point, he saw something in the dirt and knelt down to examine the ground.

Two sets of footprints, one heavy and deep, the other lighter, were clear in the dust. He looked more closely. One of the tracks appeared to be dragging a foot like some injured animal. Sam got to his feet, looked back toward the camp, and then kicked dust over the footprints. He returned to the fire with an armload of wood.

What he did not see was a solitary figure standing at the peak of a nearby hilltop. A large man studied the camp with binoculars. He wore a military field jacket and carried a rifle. But he moved this way and that, rocking back and forth, and the motion finally pulled Sam's gaze upward, and he saw the man. As he stared up at him, the man slipped into the trees on the hill and disappeared.

Sam picked up a club and slid it into his belt. He studied the sun on the horizon. Two hours of daylight left probably. He headed for the top of the hill, pushing hard up the steep incline, stumbling on rocks and branches. At one point he paused to catch his breath.

He spoke softly as if to himself. "I know you're there. I won't hurt you, but I can't help you. I'm here to die like the rest of you," he whispered. He knew he sounded like a crazy man, but who was there to hear anyway. He waited, listened. There seemed to be no one. Worse, he simply didn't expect anything. It wasn't that he did not care, but in some ways, he thought, the scenario of doom gave him a freedom he had never known.

He started up the hill again. At last he reached the summit and looked around. He could see the camp he was sharing with the girl below. Colby emerged from the tent and looked around. She heard a low whistling sound and then saw Iona perched on a camp stool where the man had obviously left her.

The little robot acknowledged Colby. "Good evening, Colby. Feeling better?"

Colby answered her uncertainly. She had never felt more disoriented. "I...don't know. Where is that guy?"

At that moment she spotted Sam slowly making his way up the hill. She saw him reach the top, look in all directions, and finally turn and wave at her. She gave him a halfhearted salute in return.

He found it noteworthy that the camp he had just discovered was directly above their camp. Sam was somewhat surprised at what he found in the camp, cooking utensils, gathered wood, an army tent, hunter's chairs, and empty beer cans. He had guessed that whoever was left would be bedfast, but they had chosen the location for its vantage point, and he was certain it had not been abandoned. He started back down to the camp.

Colby poured herself a cup of coffee and waited for Sam, attempting to calm her racing heart. If he'd meant to hurt her, he'd had his chance. She felt disgusted with herself for fainting. After all the physical, mental and psychological training she'd experienced, it was embarrassing, and it could've been the end of her quest, if he had been the enemy. The problem was, she couldn't figure out who he was. By the time he made it back into the camp, she managed a subdued "hello," but then fell silent.

As the sun dropped below the hill, she listened for the crickets. Their evening symphony should start any minute.

Sam sat down across from her. "Sorry about scaring you. I didn't even say boo."

Colby pulled her jacket tighter. Her voice was defensive. "You didn't scare me. I hadn't slept in four days," she murmured.

"Three days," Iona chimed in.

Colby was glad her face was in shadow and he could not see her profound embarrassment.

"Hope you didn't mind. I borrowed coffee from your store of food," Sam said.

Colby turned her head sharply toward her precious cache of goods, hoping that her alarm at the news wasn't obvious. "That's fine. I have enough...to last awhile," she finished, realizing how lame she sounded, how uncertain. If anyone should show confidence in some recovery, it should be her, but she couldn't rise to it, no matter how hard she tried.

"What are you listening for?" he asked.

"Crickets," she said. They usually start in about this time of night."

"You won't hear them," he said. "Maybe ever. Like you won't find the honeybees," he added.

"Oh," she said. Of course, the crickets were gone and the honeybees too.

Sam didn't dwell on the absent crickets, but returned to the topic of food. He attempted to be casual. "Your food is uncontaminated, I take it?"

"Freeze-dried and vacuum packed in Houston. Stored at Polaris depot for years. Should be fine."

He smiled at her, but not with his eyes. "How long have you been in the NASA corps? You must be the elite of the elite."

"Not so sure about the elite. Persevering, I guess. But I've been in the corps for seven years, almost two on the Polaris," she said. "I was a Mission Specialist. Agro biology. Have my doctorate from Purdue. I was on the Mars mission when they called me back."

"Building crops in space?" he said.

"We were working on the newest hybrids, and I thought we had some serious breakthroughs. We were confident we could do anything, and then the virus hit. We spent weeks of intensive study, going over the possibilities, making lists of the work we were doing that might be relevant, but nothing was. We didn't know what to do. We couldn't bring the virus up to study it. Too dangerous, the experts said."

"How so, dangerous?" he pushed.

She took her time answering. Surely he knew as much about this as she did. But maybe he didn't. She had no idea what his expertise was, but he wasn't a stranger to the disaster. How could he be, given what he had must've seen and heard at the South Pole none of which she knew. Still, telling him what she had learned seemed harmless. What difference could it make? The only place they could work would have been her father's lab and that was trashed.

"Europeans didn't take to the hybrids like Americans did. Of course, they didn't have as many mouths to feed, but the Europeans dubbed the produce 'Frankenstein food.' Over there, they didn't have agencies like our FDA and USDA telling us that we didn't have a thing to worry about, that everything was safe. The Dutch, the Scots, the Germans, never bought into that."

Sam nodded. "I've been working in the same area, but down here. I'm a crop specialist. United States Department of Agriculture." He held out his hand and smiled. "Dr. Sam Mickelson. Sorry I didn't say earlier."

"Pleased to meet you," she said, taking his hand and offering a little smile, the first sign of friendliness since they met.

Sam turned serious. "We, well, the food manufacturers, the government, we didn't really know that. A few of us questioned the premises, questioned everything, and were ignored. If you think the Antarctic gig was a reward, think again. I was conveniently out of the way. I heard I earned the title Abominable Snowman."

Colby smiled ruefully. "The agribusinesses said that Europe was simply practicing protectionism, refusing to let other's crops in. In France they even trashed a McDonald's calling them 'burger imperialists.' That was good for a laugh on American morning television."

She hesitated as if trying to remember the whole story.

"The Japanese had a little more foresight, and a little more clout. Their biggest breweries refused anything that had been messed with genetically. Wish they'd all stuck to their guns. But the powerful lobbies, the might of the U.S. Government, advertising, everything came together to destroy any opposition. The upshot of all that was that the 'new food' didn't circle the globe as quickly as the growers wished, but finally, it overcame all objections."

Sam sighed. "Yes, it would, finally. The power of the United States. True awareness came way too late," he said.

He continued. "There were some early reports that experts downplayed. I remember a headline that the corn crop in Texas showed a level of cancer agent not seen before. Back then, the corn crop was worth $400 million to the state. The carcinogen was aflatoxin, which was produced by a mold and fungus living in the kernels of corn. Drought encouraged it."

"Reminds me of some old movie when the monster, whatever it was, escaped the lab. Same thing," Colby observed.

Sam stopped talking for a moment, looking out toward the horizon, as if giving himself time to collect his thoughts as well as his emotions. Finally he picked up the thread.

"There had to be rain, the agriculture experts said or the cancer agent would thrive in the corn kernels. It aimed for the human liver, producing a cancer that was almost always fatal. Way back in 1987, a bunch of Texas dairy farmers dumped feed and milk contaminated that way."

"I had no idea about that," she said. "...although I had heard...." Her voice trailed off.

She was acutely aware that she hadn't told him about her father's work, not yet. The time wasn't right. "I didn't know about the Texas dairy farmers."

Sam continued. "The government was supposed to be monitoring the levels of the carcinogen. Their idea was if the cancer-causing entity was too high in the corn, they refused to let the farmers sell the produce into the human food chain. They'd give it to animals." Sam laughed, but there was no humor in it.

Colby made a face. "Guess they never heard of hamburgers."

"A trillion sold. What I'd give for one right now. Anyway, that was an early warning. Oh, and a story out of Mexico, that genetic contamination was found in their corn. The thing was, it was native corn."

"You're saying it wasn't hybrid corn, where you'd expect contamination?" Colby asked.

"No, the native kind where you wouldn't expect the carcinogens. The thing was, the genetic material had spread on its own. When the Mexican scientists found out, at their National Institute of Ecology, they were devastated."

"Not mad?"

"No, worse, brokenhearted. They counted it a cultural loss, because corn has always been a symbol of the Mexican people, maize, as they called it for centuries."

She looked puzzled. " They feared its purity was lost?"

"Something like that," he said. "But a lot of things had evolved. Dances. Stories. Festivals. Think about it, all that lost."

Colby remembered something else. "I know with hybrid corn, new seed had to be planted every year, since it didn't reproduce. I remember some lawsuits over natural seed. The seed industry people didn't like it when farmers sold seed that would reproduce. That cut into their profits."

"Sure did," he said.

"I see the problem," she said. "Agribusiness was all about the money."

"Yeah, but they also did the research and development. They had to have money for that," he said, without much enthusiasm. For awhile the conversation flagged. There was nothing hopeful in it, so they fell silent. It was time for their own dinner, and the freeze-dried meals were nothing to get excited about. But they provided sustenance.

Sixteen

The Seed Corn

Sitting near the fire after dinner, Colby offered more of what she knew. "There was a story out in, oh, 1999, where hybrid corn was bred to release an organic chemical to repel insects. The plan worked to perfection and once they realized it, the scientists in charge were somewhat cavalier. They said that the occurrence would benefit growers."

Sam frowned. "They didn't have a clue. Just see what happened next.'

"Now we know. It was a potential hazard."

He pushed the dirt with the toe of his shoe. "Too late." There was an awkward moment between them, as if they didn't want to talk about the elephant in the room anymore. But then Sam smiled and gestured toward Iona . "I got acquainted with your computer friend."

Iona bleeped her pleasure at being included. "Call me Iona. Everyone does," she said, sounding downright coquettish.

Sam said, "Sure will."

Colby frowned and shifted on the log where she was sitting. "Did you see anyone coming in from South America."

"Three months on the road back to Texas. A lot of bodies are all I saw. The virus has spread everywhere."

"What we heard is true then?"

"Yeah. A strain run amuck. Changed from Jekyll to Hyde. No one guessed it would circle the globe. You think you would've seen it coming? With the incredible technology on Polaris?"

"We were aware, but not to the extent that it would touch everyone."

SEVENTEEN

Warnings

Colby was able to fill in a lot that Sam didn't know. They were from different fields and had different perspectives, but their educations had many points that crossed.

She told him how the story behind the seed wars was well documented on Polaris, even beyond what she and Sam had discussed. Colby had made it her business to know as much as she could. She and Jake went over what had happened endlessly, combing through everything they received. But after every talk, they were left with the same question, which she now expressed to Sam. Why did no one see it coming.

After she had told him what she knew, he was shaking his head in puzzlement.

"For a long time, we were in the dark, but at the end of my time on the Space Station, I finally learned the truth, that there were people who suspected something." Colby frowned. "What Jake was reluctant to say was that he had an inkling about the seeds that had become commonplace in the United States. Once, when he had gone home to Decorah, Iowa, when he was almost through with his study of plant evolution with an emphasis on DNA, he was thinking about the seeds. He couldn't throw off his fears. But he still didn't really know why the evolution was occurring the way it did, with a lot of successes and no word about failures."

Sam's expression changed when she mentioned Jake. She didn't know why, but she continued sharing her experiences on Polaris.

Colby remembered Jake sometimes laughing and telling himself, and her, that it was something in his own genetic makeup or maybe it was a thing that all men did, worry about cataclysmic events and how they might save themselves and their families and yes, even their communities. How prescient he had been.

In the labs the bright minds were busy with their new hybrids, their research paid for by government grants and corporate funding, and one day he thought that maybe they would all soon forget the origins of the food system that sustained life on the earth and in the biosphere and in myriad colonies in space.

As Colby revealed more and more of what she had learned, it became clear to her that Jake did know a great deal about what was happening. She simply had not put the pieces of the puzzle together. No one had, not entirely.

"He knew that the hybrids were vulnerable to disease and sabotage, and even theft. They were without genetic memory, which left them without strength except for the imminent task of producing their crop once and only once. The means of reproducing were left to the laboratories, whether great outdoor fields waiting to be planted or secret experiments aiming to correct the flaw in the new seeds," she said.

It finally came to her that Jake had been more aware of the dangers than she thought, and that he was doing something about what he knew, and he hoped that it would make the difference in what the future held.

The one important thing Jake did do was save the true seeds until he had his own little seed bank in space. Gradually he stashed away all that was needed to begin a new crop of potatoes, wheat, rice, squash, oats and corn. No one noticed. Most would have thought him strange to bother with such an insignificant task. After all, the new seed was everywhere and according to government experts, better than the seeds of the past.

Sam found himself admiring Jake, and being a little jealous of his brilliance, but mainly he was glad he had done the things he did. Sam didn't know how the work would impact him and his new friend Colby, but he sensed that it held great importance.

EIGHTEEN

Jake's Dare

Jake Addington did not possess a shy bone in his body. In fact, so concerned was he about the threat to the earth's food chain that he called the President of the United States. He believed that he was dealing with uncertainty about the planet's food supply, including the worst possible scenario, its devastation. He knew of the catastrophes that had come before, but this one was more ominous, the ultimate threat. It could destroy life on earth in a horrific way.

At first, when he saw Colby step onto the shuttle and wave goodbye, something inside him crumbled like worn out plaster. In sequence with his household robots, he walked mindlessly through his assigned tasks. But his heart was not in space. It was with the young officer who had returned to the earth.

He remained alone for days, overseeing everything from a distance but leaving the routine chores to others. He had an excellent crew, but his absence from the bridge left him time to examine his soul, to ponder what had happened and to think about what he had hoped would take place. All of that was gone now. Was it Colby's absence, or was there someone else, his beloved Eliza perhaps, who played the major role in his deep depression.

Only when Colby called back was he able to overcome the desperate sadness that led him to self-destructive thoughts. Colby's soft voice was sad as well, and suddenly he knew he could not give up his leadership role, no matter what. At that moment, he dared himself to see it all through, whether it was providing seed for the Mars mission or to revitalize the earth. The least he could do for her was to live, for however long he had. Every minute would be devoted to discovering answers about the virus and preparing his crew for their journey onward.

Jake emerged from his self-exile. There was something about him, a look of steel in his eyes, strength in his step and in his stance. The crew recognized the powerful change in him and vowed to do their best for those unknown numbers on earth even though they themselves were outward bound.

Jake held another thought in his heart, that if he were lucky, he would have Colby back. And even if he didn't, he would know that he did everything he could to be her champion.

NINETEEN

Celebrity Opinion

Colby realized that even though they had been far apart and on different quests, both Sam and she had worked at the puzzle of the virus, asking their graduate fellows question after question in an effort to find an answer.

There were dustups all right. And there were stories that never questioned possible negative consequences from hybrids. Messing around with DNA was the business of the agribusinesses. Few seemed inclined to bother them, particularly the news media which concentrated on movie stars and politics.

Sam remembered seeing a replay of an early morning talk show in which the hosts played with the idea of the 'new foods' around 2004, years before either of them was born.

In one of her History and Technology classes, Colby had seen a copy of the same Early Show with anchorman Bryant Gumbel and CBS News Health Correspondent Dr. Emily Senay. Dr. Senay defined the "functional" or "new" foods as those that had elements added to them which made them "healthier." She added that the foods were "vitamin enhanced," and thereby healthier than organic or native plants. That irritated the organic people who'd made a huge profit off selling that line of thinking. If it was organic, it must be healthy.

Gumbel had some enhanced crackers which he waved around and started to eat, but not before he asked his questions. "Can I get the same value if I eat healthy and eat a balanced diet?"

Dr. Senay replied that it would depend on who you are and what you eat, in effect, hedging her bets. The pair went on to discuss a list of foods that had something added to them, for example, spreads and salad dressings which contained a plant extract that tasted like butter and supposedly would lower cholesterol levels. They demonstrated a dozen eggs that had "enhanced DNA," which was a fatty acid found in fish oil which was a popular food supplement that helped maintain heart and vascular health. They used the Inuit Indians as an example of a people who ate a great deal of fish and had few heart problems.

They explored the methods by which eggs ingested the fatty acid which was not natural to their development and he was told that researchers fed algae with the DNA to the chickens and obviously the DNA was absorbed into the eggs.

Gumbel was still holding the cracker in his hand. He added a little of the buttery tasting spread. He took a significant bite of the cracker and washed it down with the bottle of mineral water sitting beside his chair. Dr. Senay smiled and said that yes, the mineral water was enhanced with vitamins. She added that some products claimed to contain cancer fighting ingredients, but that had not been proven. Gumbel finished the cracker. He said that it was a little dry for him. Dr. Senay laughed. The final word was that the new foods were slightly more expensive, but as the agribusinesses increased production and investments, economy of scale should help bring the costs down.

In late 2009, Bryant Gumbel was treated for pancreatic cancer.

———— • ————

A few years before that incident, which most people had forgotten, in 2001 the usually cautious Wall Street Journal ran article after article about the "new" agriculture. One story gave a lengthy recital of how bio-engineered bugs might cure disease and aid farmers to feed the world.

Sam told Colby that if the scientists, the government and the corporations had any kind of memory, they would also recollect the failure of a bioengineered bug that had created such havoc in the 1960s and 1970s, literally destroying millions of acres of crops. Not only that, but the federal government had mandated that all cotton gins be equipped with pink bollworm burners. The result of that misguided government stupidity was that hundreds of cotton gins were burned to the ground when the burners didn't work properly.

That led to a great migration from farms to cities, with farmers and their families taking up jobs at the Ford plant in Dallas, or Oil City Iron Works in Corsicana. That job loss rippled outward, destroying thousands of others jobs as well.

The economy changed from economic solvency to nothing but land banks in a few short years. Cattle replaced cotton even in the areas where cotton had once thrived. The industrial way of life replaced much of the agricultural one in a few short years. Few bemoaned the losses of the agrarian world because the workers were much more comfortable. Working indoors in a relatively comfortable factory proved much more preferable to the uncertain and difficult life of a farmer or rancher. The women liked the change as well, because their lives became easier. It was like a late Industrial Revolution.

TWENTY

State Fair

At the State Fair of Texas in Dallas, Fletcher's Corny Dogs had been an institution for over a century. For the entire fair, people patiently waited in three long lines for the fried treat. There had been other delicacies offered up by creative fry cooks, such things as fried Coca Cola, fried peaches, deep fried Twinkies and so on, but the corn dogs remained king.

Every year, Bud Sterman made sure he bought at least two Fletcher's upon arrival and after a day of wild rides and exhibits, he usually left with a third one. His father couldn't get out of there without consuming five of those puppies. Fast food had been Bud's passion since his first Beltbuster. In middle school in far East Texas, he did a twenty-page report on Ray Kroc, founder of McDonalds, and how he turned a family burger joint into an international chain.

Bud's first job was during his sophomore year where he happily grilled wieners at The Giant Dog in Marshall. He also worked the concessions at the high school basketball games. Eventually he went to Tarlton State University and earned a degree in Business Administration, concentrating on marketing. Fast Food Management was his career of choice. Bud's plan was to hook up with a burger franchise, see how it worked, and then go out on his own.

Jack-in-the-Box in Dallas was his first stop as an assistant manager. In two years he managed his own store. Jack-in-the-Box offered him a district coordinator job, but he declined.

"Sorry, gentlemen," he told his bosses. "I'm striking out on my own."

Fletcher's Corn Dogs still had his imagination. Flipping burgers was fine, but his future was in deep frying.

Bud and his uncle, who owned a sporting goods store in Burlington, Iowa, became partners in Golden Corn Dogs, which was Bud's brainchild. His uncle provided the capital and Bud provided the know-how and elbow grease.

"It's simple, Uncle Ed. We start Golden Corn Dog over at that old bike shop on 4th Street. Then we'll expand to more locations as we grow, which we definitely will."

Uncle Ed liked the idea. He believed in his nephew and advanced him enough money to get the finest equipment. Golden Corn Dog held its own, but it did not generate enough revenue to start a second location. The problem was simple. Bud used a standard recipe. His corn dogs were no better than the ones at Sonic, Penny Whistles, or the shops in the mall.

"Maybe we should expand to burgers," Uncle Ed suggested. Bud had to consider the man's opinion.

"You're probably right. I really want to keep it simple. Do one thing and do it well." His uncle looked out at the few customers sitting outside on the picnic bench enjoying their corn dogs. People liked them. Bud was paying him a thousand a month. The place was in the black but not by much.

"Maybe you could offer a dessert. A fried pie, maybe." Bud was nodding.

"Yep. That would help. Maybe I should hire topless fry cooks." Ed laughed.

"There you go. You gotta have a gimmick." They both soberly stared at the half empty parking lot.

"A gimmick," Bud mused. "I don't have any gimmicks up my sleeve. The only thing I can think of is to build a better corn dog.'"

Ed scratched his chin. "That's not a bad idea, Bud. Why don't you experiment with different recipes?"

Bud changed meat jobbers and obtained a higher quality hot dog. He also started using a different grease product. He took his pre-mixed ingredients and added differing amounts of salt, sugar and various spices, such as chili powder and garlic. Nothing seemed to work. They simply could not push sales enough to justify opening the second locations.

Not long after, Bud was taking inventory on drink cups and mustard packets. A well-dressed young man came up to the counter/window.

"Can I help you?" Bud asked.

"Yeah, let me have a Golden Dog, a small fry and medium diet strawberry."

"Coming up."

As he dropped fries and dipped the wiener into the batter, the young man asked, "Is this your place?"

"It is. I'm chief cook, counterman and laundry service."

The young man took his numbered receipt, walked over to a picnic table and waited for his order.

When the bell went off to signal the fries were ready, Bud took the basket from the boiling oil and dumped the contents onto the metal drainer. He sprinkled a salt and pepper mix on the fries and checked on the cooking corn dog.

In a few minutes, Bud put the meal on a tray and called out the window.

"It's ready, sir!"

The young man got up and paid for the food. He went back to the bench and Bud went back to his inventory count.

As he entered a bag count on his sheet, Bud turned to see the smiling young man at the window.

"Yes, sir, do you need anything else? A fried apple pie?"

The young man offered his hand.

"Hank Cherry."

Bud brushed his hand off on his apron and shook it. "How are you, Hank? I'm Bud."

"You make a pretty mean corn dog."

"I'm glad you liked it. Tell your friends."

Hank handed him a business card. "I'm with Agra Co-op. I have something you might be interested in."

Agra Co-op. Bud had heard of the group, but he couldn't remember in what context. "What does Agra Co-op do?" he asked.

"We are one of the top genetic food producers in the world, Bud. Our guys have a new product we'd like to try. Simply put, we have a hybrid corn that releases a chemical that repels insects."

"Sounds interesting." What was Hank Cherry selling? Corn starch with insect repellent?

"Oh, it's very exciting. From that hybrid, we are working on a by-product used in the common mixes for corn dogs."

"So you want to sell me your batter."

Hank smiled. "Our Super Batter. It makes biscuits tastier and fluffier. And I think it will make your corn dogs out-of-this-world good."

Bud's mind ran over current costs and budget. "How much?" he asked.

"Nothing." Hank saw the cynicism on Bud's face. "At least not at first. We just want you to try it. When you see your profits, you won't even ask how much. I'll throw in a case of our pork links. They really are a perfect compliment to the mix."

Bud thought, why not? "Okay Hank. Send me some of your wonder batter. I'll give it a shot."

The Super Corn Meal Mix or whatever Hank Cherry called it didn't look special. There were no unusual cooking instructions. Bud shrugged as he looked at the fine granules.

"Looks simple enough." He made a batch using the pork hot dogs to complete the set. It cooked in the same amount of time, but the grease smell was replaced by a pleasing aroma. "It smells like Fletcher's Corny Dogs." He marveled to himself. When they were ready, he used his tongs to take the corn dogs out and put them on the draining rack. They not only smelled delicious, but they looked larger than the average corn dog with a sunny, golden texture.

Bud let the corn dogs cool and then picked one up. "Okay, let's eat." He took a bite. It was a flavor that put Fletcher's to shame. There was a faint, honey taste, but it wasn't too sweet. It also had the hint of butter, but not too much. Bud almost moaned as he let his taste buds absorb the flavor. And it was light. He didn't feel full. That would mean increased sales.

By the next spring, Burlington, Iowa, had three Golden Corn Dog stands. Bud couldn't get a booth at the Iowa State Fair, because they already had a jury approved corn dog. The next big event was the Iowa Stock Show and Rodeo. This two-week event attracted over a hundred thousand visitors. Golden Corn Dogs were well represented with a booth near the entrance and another in the arena along side the other food concessions.

Bud and Uncle Ed stood on the concourse watching the three long lines of people waiting to buy their corn dog.

"You did it, Bud. I'm proud of you." Bud's face was lit up as he watched the happy customers come away from the booth with two and three Golden Corn Dogs at a time.

"It doesn't get any better, Uncle Ed."

Watching the long line, Bud saw devoted customers and dollar signs. He did not realize he was also watching a funeral procession.

Twenty One

Moths And Honeybees

Colby talked about 'the big idea' out of Phoenix, Arizona, which she had learned of during her college years. Scientists had engineered male moths, millions of them in fact, and dropped them into spring cotton fields. The bugs would pass on a fatal flaw to any egg fertilized, and the result would be millions fewer pink bollworms. Or they would push out sterile male pink bollworms which was another mode of destruction.

One strain of DNA engineered plant was supposed to stunt the growth of the boll worm. It worked in a limited way, but the insect had a goal of its own, and it was difficult to contain. The work continued as late as 2011 at Texas A&M University where researchers grew their own patch of cotton and harvested it for the experiments. For many years, the scientists did not exchange ideas but worked in teams that were isolated from others with similar work in progress.

Other hopeful bioengineers worked on mosquitoes, hoping to wipe out malaria and other mosquito borne diseases.

Another "popular" pest was the Mediterranean fruit fly, an agricultural nuisance that damaged citrus and other succulent fruit crops by implanting eggs that hatched as maggots in the fruit. Particularly devastated were countries around the Mediterranean. The fear was that the pest would threaten United States agriculture.

While it was not a laughing matter except to late night comedians, something called the "kissing" bug occupied scientists at the Center for Disease Control and Yale University. They believed they could destroy *Chagas'* disease found in tropical areas and caused by the bug, an infectious protozoan called *Trypanosome cruzi*. A Brazilian doctor had discovered it and given the disease his name in the late 1800s. It was called American *trypanosomiasis* as well. Gains against it were reported in the early 21st century, but it was not completely defeated.

An unlikely creature that held the interest of scientists was the honeybee. All kinds of stories surrounded them including that they were dying in unprecedented numbers. This would then require that pollinating would have to be done by hand as it had in the past, and the cost of food would skyrocket. Many useful wild plants would soon be extinct.

Before that could happen, according to the stories that were running rampant on the new internet, the Africanized bee would overtake the domestic bee, and make honey gathering highly difficult, some said impossible. The popular term for the newcomers was "Killer Bees," which caught on instantly with the mass media. A "B" horror flick was made, based loosely on the media's visions of swarms of the bees moving north from Central America and Mexico, running amok throughout Texas and the south, particularly Miami, which didn't happen. The news hounds went on to the next big thing.

Unexpectedly, the scientists woke up. Research was thought to be so important that it was underway in the U.S., Japan and all over Europe, its goal to integrate foreign genes into bee chromosomes without destroying the native nature of the creatures. But in the twenty-first century, an Australian bee made its appearance. At first, it was welcomed as a savior for the bee industry, bringing needed "hybrid vigor" to strengthen the bees. It did not turn out well.

Alma Trainor collected samples from the rows of hives that were out in the open field beyond the live oaks. The gentle hum serenaded her as she checked each hive. Twenty-five dollars an hour, she thought. Slave wages. It didn't pay enough and she didn't like the smell, plus she'd been stung twice the last week. The protective clothes felt greasy against her skin. Hell, what was she doing here?

The BeeKeeper promised her a raise after a month's probation. That's what he insisted on being called, the BeeKeeper. The man was eighty years old, probably a younger cousin to Moses she'd read about in the bible. Bees were his life.

Alma almost regretted taking the job. She didn't have many options though. Getting fired from the kiosk at the mall had put the mark of failure on her. It was on her record, even though she had lied to get the clerk position at the power station. She held on to that job for two weeks before the manager caught her pilfering air credits off the customer roster.

All she wanted to do was get enough money to get off the Texas Gulf Coast. Maybe she could find something up Amarillo way. She'd never been there, but it had to be better. She could start a new life and she'd heard there were jobs at the nuclear waste facility where you had to wear a white safety suit at all times. One of Tom Tom's friends said he could get her a new government ID card and a driver's license. Frenchy. Yeah, he was a real identity-changer.

"I could even get you a new face, baby doll," said Frenchy. His leer told her how much he would charge for his service.

"No thanks, Frenchy, I'd rather clean my ears out with an ice pick," she returned.

Tom Tom wasn't much better. He promised her he would buy her things, take good care of her, show her the world. She had yet to leave Harris County. This wasn't long after her father had kicked her out of the house when he found all that dope. At sixteen, life didn't offer a lot of choices.

After a while, she got used to Tom Tom's ignoring her. He bought her cheap trinkets. He showed her the world of his tiny apartment in south Houston. After a while she got used to the girls he brought back to the apartment. Sometimes, she lay on the cot, smoking a joint and listened to what was going on in the next room. It always started out as drunken laughter. Then silence. Then groans of pleasure, depending on the participants. Soon the screams turned to dread and terror as the dull "thwaps" could be heard. Tom Tom liked using his fists. Slaps were not enough. And he wondered why his stable of girls didn't attract a lot of customers. Who wanted to be with a woman who had a purple swollen face and missing teeth?

Tom Tom never farmed her out, never pimped her. He just made loud, empty promises and occasionally turned sweet and made gentle love to her. Weird. This arrangement suited her for a while, but Alma knew it was only a matter of time before Tom Tom would beat her like he did the others. She had to make a change.

She saw the sign by the roadside. It was cardboard and looked like it had been written in lipstick. It read: "BEEKEEPER NEEDED - GO DOWN THE OLD SHED ROAD."

Alma walked up the long dirt road and found the BeeKeeper filling up a galvanized tub by the side of the house. His skin was translucent and his full head of hair was white as baking soda.

"Waterin' the dog," he explained, although she didn't ask.

"Yes, sir. I was here about the Beekeeper job."

He wiped his hands and stared at her though glasses thick as fog.

"I'm the BeeKeeper."

"Yes sir. You need help?"

He pointed a gnarled finger at the front porch.

"Set your bones. I will be with you directly."

Alma found a newly painted Adirondack chair. A porch swing looked inviting and as she watched the old man putter about, she figured he would prefer the chair so she sat on the swing. .

She could see past some oak trees where the hives formed rows in an open field. The BeeKeeper was slowly coming up the steps making toward the Adirondack. He fell heavily into it.

"So, you come to see the Bee Keeper," he said.

She expected a job interview, but instead sensed that a lecture was at hand as the old man took out a pipe. It was one of the main reasons she had dropped out of school. All that talk about history and math and English seemed suddenly relevant. "I came here for a job." It occurred to her she should have listened.

He looked past her. "The Chain of Nature," he announced. "That's what makes the world go round." He tamped some tobacco into the pipe. "The head bone's connected to the neck bone. You understand?"

Of course she didn't. All she wanted was employment, to get away from her boyfriend. "No, I..."

He went on as if he hadn't heard her. "The big brains engineered all kinds of things that they should've left alone."

Alma watched a hummingbird zip around the water cylinder that hung above the old man's head. She decided she could listen to a lecture if it meant a job. She noticed that the water cylinder was hanging on a loose hook. She wondered if it would come loose as the hummingbird drank. If it did, it would fall on the BeeKeeper's head. She smiled.

"Go on."

He paused a moment to think about his next statement. As he did this, he took out the pipe. Lit embers fell on his shirtfront. Alma leaned forward and quickly brushed them off. The BeeKeeper was so involved in his story, he didn't notice.

"All of this changed the course of Mother you-know-who. It messed with the food." He gave her a wink. "The big brains couldn't leave well enough alone. They worked on mosquitoes, hoping to wipe out malaria. Then there was that pesky Mediterranean fruit fly that implanted eggs which turned into maggots and that messed up a lot of citrus."

Alma stifled a yawn. Where was this going? She wondered. She was planning on a movie this evening, but it would be nice to have a job first.

"When you mess with the Chain of Nature, you can destroy the delicate balance between life and death. There were great fears in the last century that the common honeybee was doomed." Finally, he had gotten to his bees.

"Do you raise honeybees?" she asked.

He looked up at the ceiling of his porch where a wasp was attempting to build a nest. The pipe fell out of his mouth and on to his lap. Alma made a fast decision and grabbed the pipe without touching any vital organs. The BeeKeeper continued without losing a beat.

"All sorts of measures were anticipated. Some would send many wild plants to extinction." He shook his head. "Stories ran rampant that the Africanized bee would overtake the domestic bee and doom the honey industry. All of this proved out to be a lot of hogwash right out of my old granny's mouth."

He saw his pipe in Alma's hand and seemed to notice her for the first time.

"You smoke a pipe?"

She handed him his pipe. "No, sir. I was holding it for you."

He looked at it as if it was a gift. "Looks like mine."

"Go on Mr. BeeKeeper."

He tentatively put the pipe in his mouth. He took a draw on the pipe and blew smoke towards a wasp. He shook his head. "Nothing went right."

"Bad, huh?" she asked.

The BeeKeeper shook his head once more, this time with regret. "The big brains were stumped when one-third of the North American honeybees were wiped out in just a few years. Apparently, the alien bee caused a disease or defective gene or some damn thing. Who knew? The big brains certainly didn't."

Alma looked out towards the hives and wondered about the big brains. Certainly the BeeKeeper seemed like one of them. "That's terrible."

"When the bees contacted the disease, another stumper happened. In an existential effort..." He paused. "Can I say existential? Do you understand?"

"It's fine."

"Let's just say in an unusual effort, bees raced away from the hive, not in a swarm, but individually. Little honey bee carcasses were scattered for miles." He frowned. "Terrible. Just terrible," he lamented.

The BeeKeeper didn't say anything for a long while. The hummingbird left with a stomach full of sugar water. The wasp continued building its nest.

"I suppose your honey bees are okay?" She asked.

He looked up from his thoughts. "I got mine from a group in Mexico."

"That's good, I guess," she said.

"It's damn good. We need to correct the disrupted food chain."

Twenty more minutes of talk about silkworms with man-made proteins for medical purposes and how big business put money into crop genetics and then, blessedly, the Chain of Nature lecture was over.

"So what do you think, little drone?" he asked. "Does conversing with nature suit you?"

"I would like to try it, Mr. BeeKeeper."

He blew a puff of smoke into her face.

"Then let's get at her."

Now, nearly a month since she had started, Alma was collecting honey samples. The BeeKeeper had noticed some irregularities in the quality of the honeycomb. After spending years around the buzzing creatures, he could see a subtle difference in the color and texture. He was having Alma take samples to mail to Washington, to the revitalized Food and Drug Administration. She took her sample boxes to the back room of the house where the BeeKeeper kept his lab. He looked paler than normal.

"Come in, young lady. Put the box by the stove." He was cooking honey in the stove and had a slide under his microscope. The humming of the bees floated through the live oaks from outside.

"Can I help?"

"It's too late to help."

There was something in his voice that frightened her. "What do you mean?"

"Ruined. All ruined. The honey." He took the slide and held it up to the light. "Take a look." The humming sounded louder as she peered at the slide.

"May I see it under the microscope?"

"It's the color. Look at the color."

She couldn't tell the difference. "I don't...I can't tell."

The BeeKeeper took off his thick glasses.

"It's poison. You look at it under the microscope and you'll see all sorts of nasty little things swimming in it."

The buzzing was louder. Alma looked outside. A black cloud came out of the trees.

"Uh, sir?"

Alma felt first sting on the back of her neck. She slapped at it and suddenly the air vent was vomiting out a hoard of angry bees.

The old man screamed as his head disappeared under the black and gold blanket. Her arms and legs were on fire. She felt her tongue swelling. The buzzing faded.

Not everyone read about the BeeKeeper's demise, despite his prominence in the field, and no one recognized what it prognosticated. It was just the beginning of a series of ecological disasters resulting from good intentions.

Sam took it all in. He had heard some of it, but now he knew a great deal more. After awhile, the pair fell silent. He and Colby had covered a lot of conversational territory, but suddenly they seemed weary, and simply looked at one another. Something about Sam's gaze made her uncomfortable. She quickly learned she made him feel the same.

"I don't get why you're here? In this backwater town?" he asked.

She hesitated. "Family. Near...here." She did not tell him the rest, that in the morning she must go in search of her family's fate and that she feared what she would find. For now, she simply said good night and retreated to the tent.

Twenty Two

Clover

Outside in the light of a round yellow moon, Sam sat alone, staring at the silhouetted hill beyond. He had a small fire going and a cup of Colby's coffee. Fortunately she had brought several pounds of it with her and she had insisted he share it.

But he was thinking about tomorrow. Whoever was out there, he would find them. He would learn if he and the girl were truly in danger or if the few survivors of the virus faced an early death which is what he had heard at the South Pole, and he had seen the lines of bodies in Chile.

He pulled his sleeping bag off the motorcycle and laid it out beside the fire. The fire had burned down to embers, and he thought he heard the rustling of a growing wind. He hoped not, but it made no difference, he was so exhausted. He was fast asleep in minutes.

In the morning, it was raining. Thunder and lightning awakened Colby, and she had lain there awhile, enjoying the sound of the real thing. Don't let anyone tell you the faux stuff is the same, she thought, smiling and thinking of the incredible digital films of every earthly season, storms and sunshine, crashing oceans and tall snowy mountains. It was not the same. And the life-giving rain was much more pleasing than the maniacal wind that gave no peace.

Finally the storm quieted and the rain fell into a pleasant drumming pattern. It wouldn't stop her. It would take a lot more than a steady downpour to keep her from finding out the fate of her father and mother.

She reached into her duffle bag, and quickly pulled out a bright aqua slicker. She then pulled a special plastic covering over Iona who seemed to preen a little.

"Am I color-coordinated, Colby?" Iona asked.

"Yep," said her mistress. "Gray on charcoal."

Colby filled her backpack with food, a plastic water bottle, a compass and a flashlight, along with her Swiss Army knife, a durable invention that never outgrew its usefulness, even though it had added tiny flash drive portals years before as well as a phone element. She added routine things such as a shovel, matches and then emerged from the tent, surprised to see that Sam had created something like a gazebo with pieces of plank and a large plastic covering of the sort used on the backs of trucks. He had a fire blazing and coffee. He smiled up at her from his seat on her camp stool and got to his feet. "Coffee," he offered.

"Oh, thanks," she said, wondering why she was grateful for the drink. It occurred to her that he was offering her coffee he got from her supplies, but at least he had put in the sweat equity of making it.

There was an awkward moment before she said, "Ah, I have to go now."

Sam showed his dimples and the sun creases around his eyes. "Mind if I tag along?"

Colby for once was speechless. She had no real reason to refuse him. "If you've nothing better to do, join me," she said.

Sam considered the irony of her comment and grinned. "Not a blessed thing."

Traveling on a brisk west wind, almost as quickly as it had arisen, the storm moved on toward the east, where the rain would fall unheard and unfelt on places deserted of people, of livestock, of even the most primordial movement. A cold wind picked up, and the day remained gray and cloudy.

Sorry that she had not said no, she hurried out of the camp, thinking that maybe he would get the hint, but he did not. Like a friendly puppy, he chased after her down the silent main street.

As Colby strode along, Iona began a tune from the seventies, "Fifty ways to leave your lover," and played it in a half-dozen different styles. When she got to Country and Western, Colby put a stop to it. In a few more moments, they stood in front of the community cemetery which was in the church yard of a dilapidated white building with its steeple tilting like the leaning tower of Pisa, a famous Italian landmark that had fallen over in 2030 as the result of an intense earthquake.

Colby stared at the unkempt cemetery, its stones gray and pink in the early morning light, dead weeds and grass overrunning the grave sites since there was no one to take care of the place. She shivered at the chill, the abject loneliness of the place. Sam looked a little puzzled, and hung back, saying nothing. She started walking toward the base of an ancient live oak. Someone had left a long-handled shovel there.

"Grab a shovel," she said.

Sam extricated his military field shovel from his backpack. "I'll use this one. You're not planning to kill me, are you?" he asked, teasing her, but it didn't seem funny to her.

Colby gave him an exasperated look and Sam followed her deeper into the burial ground.

The girl studied the gnarled live oak and soon found a cut in it. She touched the gash, turned and calculated a distance. She then paced off four steps. "Here. Right here."

Sam hesitated. "You sure?"

"I'm certain. I buried Clover myself."

At that moment, Iona set up a howl. "Ooooh."

"Will you stow it? You are...driving...." She did not finish the sentence but picked Iona up and moved her to a spot nearby away from the burial site. Colby looked disgusted and shook her head, setting Iona's sensors in a different direction. The noise broke off immediately. "Better," said Colby.

Sam produced his shovel and pushed it easily into the sodden ground.

"What are we exhuming? A cocker spaniel?"

Iona spoke in the voice of a drama queen. "I can't look. This is barbaric."

"So, power down. You want me to dig?" she offered.

Sam stopped long enough to pull off his coat. "I need the exercise. Plus I'm dying of curiosity." He glanced at Iona with a mischievous expression.

"Oh me, oh my," said the robot.

"Iona , play us a tune. She can play anything," Colby explained to Sam who pitched shovelfuls of dirt over his shoulder, his strength obvious in his easy movement.

"C&W, please."

Iona must've forgotten how upset she was and went "mi,mi,mi. Which era? Hank Williams Sr. or Hank Williams Jr.?"

"Vince Gill style, please."

While they dug, they fell into conversation about the virus. She remembered her agronomy classes, and told Sam about the greatest agronomist of the twentieth century, a professor at the University of Minnesota, who led the way in plant genetics. The hope was that the hybrids heralded the most productive period in the history of agriculture. The world would be fed, she reported, repeating what she had been told. Crop failures would be a thing of the past. It might've worked, until some realized food could be used as a weapon. That was the downfall of all the fine projections of feeding the world. It would happen only if the people followed the leaders who had turned greed into a positive, if only for themselves.

Iona played only a few bars of a Garth Brooks song when the sound of metal hitting rock made them all freeze.

The robot reverted to her injured state. "It's too awful to think about."

Out of nowhere, Iona set up a howl. "Ooooh."

Colby looked disgusted and shook her head, settling Iona nearby. "You must've been programmed at a girls' finishing school in the nineteenth century. A little more guts, please."

There was a moment of silence and then a small voice piped up: "What are guts?"

"Never mind. It's not a concern of yours," Colby said. "I put in a layer of rocks."

Sam kept digging and Colby joined him. Colby's shovel scraped metal. She eagerly began pushing aside the dirt. Sam stepped back when she dropped to her knees and caught hold of whatever was down there and pulled it out. She tore off a layer of plastic and revealed a ruined earlier version of an IONA. It had a Roman numeral I etched on its label which had almost disappeared.

"Ugh! Let sleeping computers lie."

"An old computer? And it's been cannibalized." Sam said, a puzzled expression on his face.

Colby brushed off the rusted computer and smiled fondly. "This is Clover, my first computer." She fiddled with the buttons.

Sam took the machine and examined it. "Dead as a rock."

Colby was disappointed at first but then resolved. "Dad sent me here. There's a reason." She leaned over the hole and began to dig in the "grave." Her reward was a small plastic container. She tugged at it and then triumphantly held up a miniature flash drive like the one she used with Iona. "This is the reason we're here."

Sam watched her expectantly, but she picked up Iona and strode away. He started to follow, but her look waved him off. It was a definite "no."

Colby went a few paces more and stopped beside a towering tombstone that appeared to be among the oldest in the burying ground. She placed the information device into Iona. "Play this."

Colby listened expectantly, but Iona seemed not to be focusing on her task. "Why is Sam down the hill?"

Colby sounded like an elementary school teacher. "Pay attention, will you!"

Iona's sound changed then, to the masculine voice of Colby's father. "When you hear this, you begin your search for the life source. You will find it, if you follow my instructions. Everything depends on you now. Go to the Blue Room. A song we sang is there. Remember, trust no one."

Colby jerked her head toward Sam who was standing outside the cemetery, leaning against a tree.

"Is that all?" she asked.

"That is all."

"The blue room?" Colby repeated.

Twenty Three

The Watchers

Colby did not see two aged figures of Angel and David Sloan hiding in a grove of trees about a hundred yards away, and up the hill. Another figure, a woman named Wanda Toten, lay on an Army cot behind them. She was propped up on knapsacks, struggling to breathe. The tallest of the men, David, stared down at the camp site, rubbing his left arm, which had gone to sleep while they were surveying the camp below.

Angel adjusted his binoculars to a higher power, and then took them away from his eyes. "She'll get on with the search now," he said.

"We never should have left them there," whispered David. He looked old as Methuselah now, with deep wrinkles that had appeared within the past week. He had lost height as well, as his spinal cord tightened and became more compressed. It was harder for him to accept the changes in his body, because he had been an athlete. He had played baseball and got a couple of years on a minor league team roster. On one occasion, he'd been called up to the Texas Rangers in Arlington, but it was as a pinch hitter and he didn't stay long even though they had won the World Series. It bought him a good many cups of coffee though, since he had batted against a series of New York Yankee pitchers. He had run the triathlon not so many years before. Only a few months ago, he had realized that his running days were finished. At the time, he attributed his decline to the passing years. He had no idea that his body's aging had gone into overdrive. It was that memory of what he had been that kept him going. He seemed to mention it more often. Angel, who had played soccer for Rice High School, was angry, his voice sharp as he turned away. "That's all past. We've no time to dwell on what is past!"

David kicked at the rocks beneath their feet. "Do you still think Sam will help us, with her here?"

Anger rose in Angel. "He must! I want you to go back to the farm, give the old man a heads-up. I'll be along shortly. I'm going to stay with Wanda. I don't think she'll last much longer." His voice trailed off as he stared over at the dying nurse and went to sit beside her.

David obliged him and headed toward the farm. Angel waved goodbye and then went to her. He reached out with his own weathered hand and held Wanda's weakened one.

She smiled up at him and began to whisper.

"What...what are you saying?' he asked.

"Is school out?" she asked.

He smiled. "Oh, yes, it's out. The children are playing."

"Good, children should play," she said. "I tried to warn them. The virus would take us all, but no one heard." Her voice fell silent.

He didn't have long to wait. Like everyone else he had seen die, his parents, his friends, his fellow workers, she simply closed her eyes and took one last breath. By now he had no tears left. He covered her body with an Army blanket and stepped away. Tonight, they would return and bury her, if they could find the strength. At least the backhoe was still operational.

In a little while he made his way to the old community cemetery. The young people were gone now and he wanted to pay his respects to his father, Jorge. In a few minutes, he stood beside the grave and stared at the lettering on the tombstone. He knelt down, not in reverence, but because he was exhausted. What Angel knew and what the others would not mention was that their time was limited. Was there any hope? Only if the girl Colby had brought something to reverse the aging. Or if Sam had found an answer. Awkwardly, Angel, the once young man who had been the best shot in the county, got to his feet and leaned on his rifle. He knew that he should prepare for the inevitable, but what he did not know was that he had Alzheimer's. Something had worn out in his brain, and it made him want to hit out at someone, anyone, even the young people.

Twenty Four

The Town

The place probably looked the same fifty years ago, except that the streets were deserted now. The chill wind had picked up and was blowing fiercely. The rain would come back, that was certain. Cars of every description lined the streets, debris as if from a parade piled against flattened tires.

Sam ambled along behind Colby, catching her as she paused in front of a store and sat down on an old bench. For one instant, she put her head into her hands and leaned down, as if she had given in to the terrible aloneness, but then she sat upright, as if in her best military form, and picked up Iona and began to speak. "Arrived Rice, Texas, November 4, 2042. Home. Final leg from Houston Space Center required four days. Fuel problem solved with fuel cells left in stations along the way. Saw no one alive. Except here in the town I found Sam Mickelson."

Sam waved his hand at her and pointed to himself. "I found you," he whispered.

She ignored him and continued her report, although what good it did she didn't have any idea. Habit, she thought.

"I don't know who he is or what he's doing here. Our fears are true. Earth devastated by agricultural disaster. I will pin down details. I am now following my father's instructions. I have no other choice. Got it, Iona?"

Iona made her usual whirring sound to indicate that she had recorded and stored the information. Colby got up from the bench and continued along the deserted street. On a store window were giant painted Panther Paws and the slogan FIGHT PANTHERS FIGHT. Colby stopped and listened to the wind howling, and for a single moment, imagined she could hear the distant, muted sounds of a high school football game and all of the people yelling for the home team. If only it were true.

Colby turned away from the window. Her feet were frozen and her body refused to move. Why should she? It all would end in the same disaster she was seeing here, once the food ran out, and with Sam here, it would run out sooner than she had thought.

The sibilant sound of the motor cycle engine caused her to pivot, and there was Sam, doing a wheelie. He stood it almost on end, and then raced toward her, coming to a screeching stop near her.

"I forgot to give you back your property yesterday." He tossed the canteen to her and she caught it, a little irritated but it took her mind away from the worst of the possibilities that wouldn't stop haunting her. "I filled it up."

Colby frowned. "With what?

"Water, of course, silly. From your supply. Come on."

The idea of his taking liberties with her supplies troubled her, but only for a moment. Colby stepped out into the street. At least somebody was cheerful. She hesitated, and then raised her hands in mock surrender. Why not, she thought, and climbed onto the back of the cycle. Sam grinned, gunned it, and off they went.

TWENTY FIVE

Home, Where You Can Always Go

The motorcycle halted on a residential street. The neighborhood was once a pleasant place, its streets lined with tall cottonwood and live oak trees. It was autumn now, and their leaves had drifted to the ground for the most part. The street itself was long abandoned, and the windows of the houses were like great dark eyes staring out of cold empty bodies. The yards showed the months of neglect like orphans abandoned. Colby hopped off the cycle and started boldly for the house, but then slowed her step. What was she so eager to find?

Sam stood the cycle up and followed, his hands stuffed in his pocket like a kid on a picnic. His eyes belied his look of physical ease, moving back and forth as if he were tracking enemies at three and nine o'clock. He whistled loudly, until Colby, irritated, turned and gave him a look. He stopped whistling abruptly.

"This is it," she said. "This is where I was coming."

Sam kicked at a branch that had fallen onto the sidewalk. "Whatever you're looking for, you can forget it. Everyone's gone. Everything's gone."

For the first time, she detected a trace of bitterness in his voice. "I'm not giving up yet."

Sam shrugged as if she was nuts. "You going in there?"

"It's why I came."

She started in and glanced behind her. He was trailing her, listening to every word.

"Alone."

"I'll nose around out here," he said, his voice agreeable.

Colby watched Sam disappear around the side of the house, and then slowly, she pulled open the noisy rusty-hinged door and went inside.

Across the street, the gray haired man was kneeling behind a deteriorating brick wall, tall trees hiding him from Sam's view. Could Sam have seen him, he would know that it was the same figure that had been watching the two young people since they had come to Rice. He stared at them intently, his eyes almost too dim to follow their movements. The man used his rifle to balance himself in his crouched position. He would wait.

TWENTY SIX

Last Things

Colby started through the dusty hallway. She straightened a wall hanging and moved on, finally stopping before a door. Her hand darted out and she jerked it open hurriedly as if waiting would halt her progress.

Sam found himself at the back of the house. He looked into one old building and saw nothing. But then he spotted a small modern outbuilding. He tried the door. Nothing. He stepped back and kicked in the door. His hand went involuntarily to his nose, and then he went inside.

Inside the house, Colby discovered a bedroom that was left neat, although the layers of dust suggested it had been abandoned. A blind had fallen down, but there were still framed family pictures, an old-fashioned four-poster, an easy chair and several familiar college diplomas.

She picked up a photograph from a chest of drawers and stared at the familiar faces of her parents. She laid a finger on the photo as if by doing that she could touch them. The expression on her face was painful. After a moment, she put the picture back with the others and backed out of the room, softly closing the door behind her.

At the end of the long hallway was another door. Colby walked determinedly toward it and went inside.

The playroom had been painted to a near navy shades of blue, and a lighter pallet suggested orderly almost military choices. Though dusty it still had not changed. The walls were lined with photographs of her, a mini shrine: playing the piano, wearing a dance costume, reaching for a diploma, standing in front of the Naval Academy with suitcases and her parents. In later photos, she was dressed in fatigues, and there almost life-sized, Colby attired in the uniform of a U.S. Astronaut.

She gave herself only a moment to think about all that had been, and then she began moving resolutely around the room in search of something. At last she began to wind up a ballerina music box. The tune was louder than she had expected, a Chopin piano concerto. Quickly she closed the box.

Iona piped up. "What exactly are we seeking, Colby?"

"Music, maybe. I don't know," she said.

"I like songs with words."

Colby was irritated. "When I find the song, the words are supposed to tell us something."

Colby walked to the dusty upright piano, placed Iona on top and flipped through layers of sheet music.

Colby sounded impatient. "Dad didn't make this easy, did he?"

"If it were easy, anyone could find the life source...and destroy it," said Iona, sounding maddeningly like Hal the computer in the old movie "2001: A Space Odyssey," a favorite on Polaris.

"I know that. I'm just frustrated." She picked up a piece of music and studied it. Unpracticed hands began to play the notes, timidly at first, and then more forcefully. The few bars of jazz somehow made her feel better. *Skylark* reminded her of another tune that she knew by heart, *I'll Be Seeing You.*

Iona began to hum it softly. "'I'll be seeing you in all the old familiar places.' That's a rare old song!" At that moment, the beautiful voice of a woman picked up the song. "I'll be seeing you, in all the old familiar places, the carousel...the small cafe, the park across the way, the wishing well."

Colby held out her hand for Iona to stop. "Remember, this is a clue. What is it telling us?"

Colby took a step one way and then turned around, something she had often done while studying or attempting to solve a puzzle.

"The well. The wishing well. In the center of the town. Remember! You're a smart little automatic."

"I was nicely programmed."

With that, Colby picked up Iona and raced for the door. Outside, Colby looked around for Sam, excited at her discovery.

Sam stepped out of the building where he'd kicked in the door. He saw her coming and closed his eyes, turning away from her, pretending to examine something nearby. It was no good. When she saw his face, the words died on her lips. She walked over to him, a question on her face.

"I found what I was looking for. What's in there? What did you find?" Of course, she knew already.

Sam looked off in the distance, his usual smile absent. "There are..."

Colby held up her hand to stop his speaking. She started for the door. Sam restrained her for a moment.

"You don't have to. I'll take care of them," he said.

She looked at him, a strange fixed expression in her eyes. She took a step toward the door. "I do have to." She pushed past him, and went inside. Sam ducked his head and leaned against the building. In a moment, Colby stepped back outside, her face white. She walked a few steps away and began to spit out the unbidden bile that surged into her throat, burning like acid.

Sam waited until she had coughed and wiped her face and turned back to him. "We'll bury them."

Colby nodded in agreement. She refused to say it, but she was relieved that he was there to help her.

In space, they were a young corps of astronauts and had few dead, and when one did die, of an accident usually, the body was incinerated in minutes and the ashes stored in a columbarium or given to dark space in a long trail of gray against the blue black surrounding them. This then was something different. Sam had been kind enough to encase the remains, skeletons, in plastic bags and then body bags they had found at the fire station. He had even taken them to the grave site she had chosen, pulling them in a small cart attached to her vehicle. Then he dug the graves himself.

She had thought she would engage the "sliver of ice" in the act of burying her parents, but at the end, something unexpected touched her. These were her parents, laid to their final rest, by her friend. As Sam poured the last shovelful of dirt into the fresh hollow in the earth, she dropped to her knees unbidden and remembered the words so often repeated by others in chapel. "Our Father, who art in heaven, hallowed be thy name." The words were both new and comforting, coming from her own mouth.

Sam had joined her, and at the end, reached out his hand to touch her, only she had pulled it away. "Amen," said his strong voice. "Amen," she said.

Twenty Seven

The Eminent Doctor Ryder

Maurice Ryder was one of the most gifted students ever to study at the Massachusetts Institute of Technology, a storied institution known by many as just MIT. It was founded in 1861 and located in Cambridge off the Charles River. He had gone there as a boy of seventeen, and when he emerged seven years later, not only did he hold a PhD in plant viruses, but he had completed his degree in medicine as well. He was one of a half-dozen brilliant young people who was also studying viruses in humans. Beyond that, he was no longer a student. He was the teacher now.

When he talked others listened. They would sit mesmerized at Dr. Ryder's careful choice of words, his ideas far ahead of most. The students, and even a few other professors, did not want his lectures to end, mainly because the lectures were optimistic, expressing an unlimited future with such promise. The new crops he continued to study would feed the world, he was convinced.

This was a university where lectures generated as much excitement as a pep rally or homecoming. Everyone wanted to be in on the revolutionary ideas and theories that sprang from the fertile minds of great educators. Ideas that changed the world. If there was a rumor that a particular professor was going to speak, an enterprising student could rent the lecture hall and sell tickets. Well, almost.

This was the case every time Maurice Ryder held court in his lecture hall. Three classes a day with a hundred and fifty full seats, Dr. Ryder never failed to satisfy the curious minds. After his graduate studies at MIT, he was immediately offered positions in both science and med schools.

Dr. Ryder opened the fall semester with a lecture on genetic engineering in foods. He addressed the hundred and fifty students by pulling out a McDonald's quarter pounder with cheese.

"Would anyone like a bite?" he asked.

This was greeted with appreciative laughter. He then pulled out a large fry and medium drink and set it on his desk. "Well, now. We have a complete meal here. I can read the fat and cholesterol contents on the box here." He made a show of reading the box. "Ooh, scary." This elicited more laughter. "We are very calorie conscious today. We are also cancer conscious. AIDS conscious, autism conscious, weight conscious and heart conscious. I could go on."

He held the burger up to his mouth.

"Remember, kids, I'm a trained professional, so don't try this at home."

The class exploded with laughter as he took a bite and munched contentedly on the morsel. He swallowed, took a sip of his drink and wiped his mouth. "One benefit is, this burger does my endorphins good. I eat one of these at the beginning of every work week. It helps me get past Monday." He took advantage of the next wave of laughter and popped a French fry into his mouth. Then he took a swallow of his cola. "That's right. We all go to McDonald's because it tastes good. We don't go there for our health."

A young woman on the front row raised her hand.

"Yes, Glenda?"

"I'm a vegan, Professor Ryder."

"Then you must be a very unhappy young lady."

Glenda laughed along with the others as Maurice Ryder took another bite. He chewed and swallowed. "I've been telling the McDonald's people for years that if they offered a veggie burger, they would generate billions of more dollars." He took another swallow of drink. "Everyone go to lesson seven."

The room suddenly sounded like it was being invaded by crickets as fingers clicked across the computer keyboards.

"You will see a table of ingredients I devised that will form the future McDonald's burger. The near future." A young man behind Glenda held up his hand. Maurice wasn't sure of his name. He thought it was Robert.

"Yes, Robert."

"Robbins, sir. Sean Robbins."

"I'm sorry, go ahead, Sean."

"I noticed on your list here you have an angora enzyme listed."

"Very good, Sean. And no, that doesn't mean we'll be eating Big Mac's made of mutton. We're talking cross breeding with radical enzyme and growth hormones."

Scuds Malloy, who sat on the front row, always had a comment. It was rarely relevant to the class discussion.

"Dr. Ryder, are you advocating this voodoo science?"

Ryder suppressed a smile. Gestured to Scuds to expand his commentary.

Scuds half turned to the class, but spoke to Maurice. "You said they were experimenting with bovine ovaries to enhance reproduction in younger females."

Maurice rolled his eyes. Scuds sometimes made appearances at a local comedy club in Boston. His material was suffering, because he wasn't getting laughs from the class. "So now you've got little lambs giving birth to little lambs which make really tender meat."

Scuds was so serious with his delivery, even Professor Ryder was beginning to believe his nonsense. Glenda, Sean and the others were shaking their heads. They'd seen his act before.

"What does that tell you, Scuds?"

"It means when I eat my chicken tenders, they'll melt in my mouth."

If that was his punch line, it was met with stony silence. Maurice finished the burger and wiped his mouth.

"That's the first time I ever ate a visual aid." This brought the house down as the students roared with laughter. Even Scuds suppressed a smile and made a note to remember the line. Dr. Ryder turned serious. "Scuds was being facetious, about the bovine ovaries, but he's not far off on the absurdities that are performed in some labs. On one side of the chart, you'll see I've included a solid list of ingredients that could form a healthy, flavorful burger. The other side illustrates the methods some use to push the envelope on research." He looked over at Scuds. "Ovarian eggs fertilized and growth chemicals. That's what I said. In some quarters, there are no rules. Anything goes. This can be dangerous. I know it sounds crazy, but if certain trends continue, an entire species could be wiped out."

Scuds spoke as he held up his hand to be called on.

"Plus, the greedy corporations want to make a bigger, cheaper burger and soak us poor anti-vegan saps." This actually got a few twitters.

Professor Ryder addressed the class. "Scuds is more right than he knows. I like capitalism when it's wrapped around an honest dollar, but the bottom line is money. And unfortunately, that's all our government and some conglomerates care about. They do not consider the risks... the side effects...the consequences."

On their computer screens, the students saw a design of a seed in various stages.

"I was using the burger as an example to illustrate cross genetics in the food chain. Specifically the seed virus that could change our lives drastically." This was met with blank and puzzled expressions.

"What seed virus, Dr. Ryder?"

"Right now, it's just a theory based on what I've seen happening. Scroll to the next page and you'll see what I'm talking about."

For the next hour, Dr. Ryder spoke of the food shortage that could result by cannibalizing natural earth grains. He painted three possible scenarios from unhealthy, pseudo-foods to worldwide famine. As he finished his discourse, Dr. Ryder gave them a warning.

"Remember cause and effect. Many of you will go into the research fields. You will be seduced by the exotic plant cultures developed by gamma rays and the "breakthrough" genetic engineering of species. Just remember that the simple, God-made bean is perfect in every way. There's nothing wrong with improving natural foods as long as you test, test, and test some more. Be sure. Challenge your own theories, test your products and test them again before you unleash them on mankind."

Amid enthusiastic applause, Maurice took his empty French fry box and burger wrapper and stuffed them in the McDonald's sack. He grabbed the remains of his drink and his laptop and headed out.

———

The Student Center was a beehive of activity as students left classes and headed for the bookstore, cafeteria or game rooms. A few students greeted Maurice as he made his way to the lounge area, but he looked around and noticed that most of the students had their noses buried in various electronic devices. He sometimes wished they would look up and around them and interact with other students, but of course, the other students weren't doing that.

"Afternoon, Professor Ryder," chirped a blond, blue-eyed co-ed. He nodded his head and returned the greeting, but his eyes were trained on the red-haired beauty lounging on the couch in front of the wide-screen TV. She caught his eye and smiled. She made a show of getting up, but he waved her down.

"Stay there," he commanded amiably.

In her ninth month, Linda Ryder gratefully obeyed.

He sat beside her and planted a kiss on her cheek. "Hello, sweetheart. Have you been waiting long?" he asked.

"I had a student conference and it ran over. I thought I'd be late." She noticed the wadded McDonald's sack and touched her belly that was full of life. "We're not raising her on quarter pounders."

He shrugged innocently. "Hey, this was a visual aid."

"No, it's your Monday afternoon pick-me-up."

He put on a face full of shock. "How did you know about my personal hell?"

"Students talk."

"Hmmm."

Linda Ryder had an undergraduate degree in psychology and a brand new PhD in Applied Psychology. She had the duel role of teaching classes and running a small, but exclusive practice in Boston. Despite her condition, she had continued teaching although taking a lighter load this semester.

They met when Maurice was in pre-med. Linda, a freshman, was unsure of which field she would pursue. On the particular day they met, she was in the midst of her cross-country training, something about which she could make a decision. Displaying athletic talent, Linda Cook participated as a University swim team relay swimmer as well as track participant. She ran cross-country. It was rumored that she had caught the attention of Leslie Colburn, coach of the Olympic team.

Linda was pushing herself, hoping to convert her cross-country skills into the Boston Marathon. She had been running all through Cambridge and now had returned to campus, morphing into an all-out sprint. Keeping her eyes on the imaginary finish line in front of the administration building, she didn't see the half-flat soda can in her path. Her foot caught the uncrushed part of the can and she twisted her ankle badly.

"Oof!" She fell and skidded on the concrete, scraping the palms of her hands and her knees. It was the ankle pain that took her mind off the bloody ridges on her body. "Oooh!" she moaned, rolling in agony. Several students and a professor gathered around her.

"Are you okay?" asked a young woman who wore an MIT sweatshirt.

Another student dropped his books. "Let me take a look at that."

"Careful, it hurts." It was her right ankle and it was swelling up like a balloon attached to a can of helium.

He gingerly handled the injured ankle as he examined it. "Don't worry, it's not broken. It's just a bad sprain."

Tears of pain and disappointment were making their appearance. The sprain, she could take. Her main thought was on her lost opportunity for the Marathon.

"Are you sure?" she asked, hoping that he knew what he was talking about.

He nodded confidently. "You need lots and lots of ice, constant leg elevation and some of that clam chowder they serve over at The Fish Hook."

She realized that all of the other people who had gathered around her were gone. Linda turned her attention to the Samaritan's handsome face.

"Are you asking me out?"

"Not really. I would have to carry you all the way over there."

She noticed his thin shoulders. He didn't look very strong. Then he laughed.

"Stay there." He left his books with her and sprinted off towards the nearest building.

She watched his receding figure. "Is he coming back?" she asked no one in particular. That's when she saw his books scattered at her feet. She read off the titles. "Hmm. *Grey's Anatomy, Horticulture for the 21st Century, Chromatic Dialogues, The Viral Nation*. Must be a theatre major." She laughed at her own joke. The sudden movement shot pain from her ankle to her groin. "Ooh, note to self, be still."

She saw him come out of the admin building taking two steps at a time. He was carrying something blue. He rushed up to her. It was a blue cloth napkin with ice.

"They installed an ice maker by the cola machine. They like their Pepsi in Styrofoam." He gently laid the compress on her ankle.

"It's cold."

"Hang in there. Let's sit here a while, until your swelling goes down." She watched his face as he carefully repositioned his concoction.

"I understand the ice, but where did you get the napkin?"

He jerked a thumb back to the admin building.

"Dean Dexter. He has silverware, fine china and linen napkins in his office. I once had lunch with him. He eats like some despot king." He laughed. "I think he makes his secretary wash the dishes."

"Does she have a dishwasher in the office?"

"He, not she. I think he has to take the plates home to wash."

"There's a harassment suit in there somewhere."

"You must be pre-law."

She nodded to the books. "You must be pre-something scientifically technical."

"Pre-med." He picked up *The Viral Nation*. My passion. I plan on having a degree. How about you?" He noticed her firm, hard calves and swimmer's shoulders. "Wait, let me guess. Volleyball 101 and a half."

She looked at him seriously. "You are exactly right. How did you know I was delving into the jock sciences?"

"I'm talented that way." He offered his hand. "I'm Maurice Ryder."

"Do people call you Maury?"

"Only this kid who bullied me in the eighth grade." His eyes looked distant. "Wonder what happened to her?"

Linda laughed. "You're funny."

"I'm funny and you are?"

She extended her hand. "Linda Cook. And I'm not a jock primarily. I'm taking psych courses. She wasn't quite ready to reveal who she was, not yet."

They spoke for a few more minutes, probing for common interests and learning of their roles at MIT. Finally, Maurice stood.

"Can you get up?"

She took hold of his arm and tried to put weight on the ankle. She grimaced.

"Put all of your weight on me."

She was his height and felt if there was a fistfight, she could take him. She was surprised at how hard his muscles felt. "Do you work out?" she asked.

"My life never works out," he said flippantly.

She smiled. "Maybe your luck is about to change."

Now, sitting in the lounge area, Maurice eyed his wife's pre-birth girth.

"I made up a song for her."

"Pardon?"

"I wrote it so we could sing it together as she grows up. It will be our song."

Linda cocked her head and squinted at him. "Are you serious?"

"What better way to bond a father and daughter than to have a song."

"Yeah, my father tried that with me. Old Beatles' songs. It didn't take."

"This one will." He looked at Linda's swollen belly and sang to it. "If you're feeling really low, really low, really low, if you're feeling really low, sing this ditty."

Linda was frowning. "Maurice, that's 'London Bridge'."

"What?"

"The tune. It's an old song, London Bridge. You're plagiarizing."

"From whom?" he challenged.

Linda looked puzzled as she searched for a name. "I don't know. Maybe you stole it from some guy named 'Traditional'."

He sang the second verse to her stomach. "When you find an empty box, fill your socks, fill your socks. When you find an empty box, sing this ditty."

People were stopping and watching Maurice sing to Linda's stomach. She took his hand. "Ahmm, Maurice..."

Oblivious of his audience, he sang the last verse. "If you feel down, just look up, just look up, just look up. If you feel down, just look up, and end this ditty." He looked up, startled at the applause from the onlookers. He recognized Glenda from his class.

"You should have been a composer, Dr. Ryder." Maurice stood and took a mock bow. Linda turned red with embarrassment. She waved off the others.

"Just move along, folks, there's nothing to hear, here."

The group began to disperse.

"What did you think?" he asked.

"What does that song mean?"

He shrugged. "It doesn't mean anything. It's just a song."

She leaned over and kissed him.

"You are a sweet man."

"Are you sure you feel like a movie?"

"I feel like swimming the English Channel. Besides, it's Brad Pitt."

"I don't know what you see in him."

"That's because you have too many male chromosomes."

Linda was due in a week, but he'd been expecting her water to break for a month. He stood and as he did when they first met, he offered her his arm.

"Wish I had you when I sank into that leather chair in my office," she said.

He got her up and they headed for the faculty parking lot. Linda touched his arm and stopped. "Hey, that's an idea."

"What?"

"A chair for pregnant women. You sit and sink into a nice, comfortable chair. Then, when you're ready to get up, you hit a button and it pops you out."

"You mean like a pilot ejecting from his jet?"

"No, it would be a nice, gentle rise with solid back support."

He patted her hand. "You just keep thinking, Butch, that's what you're good at," he said, quoting a 20th century film, *Butch Cassidy and the Sundance Kid.*

On the big screen, a young Brad Pitt as Billy "Doom" McCall, ran over the Paris rooftops in his tuxedo. The terrorists were in hot pursuit as usual. They often were in 2012 when the film was made. When he came to the last roof, he spoke into his French cuffs.

"They're on to me."

A helicopter suddenly appeared and a rope ladder dropped in front of him. "Just in time." he said, as he shimmied up the ladder. When he got to the cockpit, he gave the pilot a thumbs up. "Good job, Bruno. Let's get out of here."

As the copter swooped up, out of the range of the terrorists' bullets, Bruno lifted the helmet visor. "Good to see you, Doom." Anne Hathaway, playing Captain Bruno Blue, gave him a come-hither smile.

Pitt wagged his finger at her. "No time for that, Bruno. You've got to get me to Monaco before that bomb explodes." He gave her a wink.

Maurice shot a sidelong glance at Linda who was prone to moan every time Pitt got flirty. She was quietly watching the screen, munching popcorn.

The screen turned bright as the helicopter left the darkness of Paris to early morning in southern France. Standing at the dock in Monaco was the main villain played by Jeremy Irons.

"It's too late, Doom," he said, watching the helicopter draw near. He pressed a button and there was an explosion.

The audience gasped. Maurice felt Linda clutch his wrist. He looked over at her and she was smiling.

"It's time."

An hour later, Colby Aurora Ryder made her appearance in the world. It was 8:08 p.m., August 2nd, 2014.

———

The federal government kept watch for such young men and women. The National Security Agency recruited him even before a teaching hospital could. No longer was it considered appropriate for the United States government to shake and rattle its nuclear weapons. Now it was considered food, which itself was a kind of weapon. Hadn't nations used sacks of rice and grain as tools for gaining young men and women to their ranks?

In the U.S. such enlistment simply paid a great deal more. The thing that it did include was the chance for Maurice to continue his study, which he did. Before he was thirty, he owned a second PhD, and a new wife who was herself a scientist, Dr. Linda Ryder. While his specialty was viruses in basic biofoods, hers was psychology. Pretty soon, the scientists realized that their beautiful child, Colby, inherited the family brains as well.

Before the killer virus hit, Dr. Ryder had been surprised to be sent to Central Texas. His assignment was to study a strange corn blight that had destroyed the plant over a two county area. His wife decided to take a sabbatical and went with him. Their daughter, reluctant to leave the East coast, accompanied them for only a short time. She made no effort to make friends in Rice. Her eyes were on the international space program, an ambition her parents encouraged.

Meanwhile, Dr. Ryder was invited to speak at every school, club and church in the heart of Texas. His message was positive. He was well-received, and when any questions came up about viruses in seed or food, he was the go-to guy. He was quoted in the local social media time and again. No one would have guessed that his admirers would turn against him when the severity of the virus became apparent.

On a morning in August, 2040, Maurice Ryder walked back into his makeshift laboratory and tried to adjust the digital microscope. He had abandoned the effort a few minutes early, totally frustrated at what he was seeing. He thought at first that his eyes were playing tricks on him. He knew he had the beginnings of a cataract, of course, and he had a trip to Dallas scheduled to repair that. No, it was not his bad eye.

The DNA evidence he was studying had changed quite literally overnight. Now it was unrecognizable. Such a thing was impossible, but here it was. He had to make a decision. Was it time to tell his superiors? Or should he give it more time? He moved restlessly back and forth in the laboratory, weighing the possibilities.

"What are you up to?" the voice of his wife Linda startled him. He looked at her sympathetic face.

"Deciding. Maybe you can help me," he said.

They took coffee and went out onto the small deck at the back of the portable building which was his office, and he began to tell her his concerns.

Twenty Eight

Sam Before

He had always been a hero, in high school, in college at the Naval Academy, later in graduate school, and even astronaut training which he eventually dropped out of when he realized he was claustrophobic. Still everyone looked to him for leadership, and he did not fail to provide it. Along the way, he met Eliza Addington, who was, according to Sam and other young men he knew, the "perfect girl." Even his mother thought so.

The one caveat Mrs. Mickelson had was that Eliza possessed too much ambition for her profession. For as much as Sam wanted to travel into space, she desired it more and was willing to give up everything else to achieve the goal.

He wasn't listening to his mother, of course. Instead he remembered the first day he met the girl, at the Naval Academy their senior year.

——◆——

The activity lounge at the Academy was an acre of navy blue carpet with equipment that challenged every muscle in the body. Sam hoped he could work off some of his anxiety as he mounted the trotter. He'd been told he was not spaceflight material, which had gone through him like an electric shock.

Trotters were like the old treadmills. It self-adjusted the speed as it monitored the capabilities of the heart and leg muscles. The only choice Sam had was where he wanted to go on the trotter. The stationary track had a screen in front. Sam donned the special glasses and chose a popular jog path in northern Colorado. As the track sped up, the screen revealed a rustic path. He was running down a slope with aspens on each side. The beauty of the Rockies showed in the distance.

The illusion of jogging in Colorado could not take the worry out of Sam. As he high-stepped up some rocks, his fear of Jake Addington grew. He was afraid the Commander didn't like him. He was more afraid he would not choose him for the assignment that he wanted. If the Old Man didn't pick him, it would mean he had his head in the wrong place, like up his butt. Sam had been so sure he was right for the job, that when he failed the test for suitability for space travel, he could hardly believe it. After all, he always succeeded at what he tried, until now. He still believed in himself.

His lungs began to burn and Sam tried to push it, but the trotter knew better. The track slowed down as it monitored his heart rate and brain function. The woods of Colorado disappeared from the screen. Sam had accomplished the optimum workout.

He took off the glasses and saw what he thought was a celebrity in an isometric booth. The young woman was pulling her arms over her head, then back. An invisible force created a gravity atmosphere heavier than that on earth. People stood in the booth and moved around, fighting the unseen force. It was like being in water, except you could breathe. The isometric booth was more effective and safer than the old-fashioned weights that Sam had back at his apartment.

The young lady's body was well-toned and her tan was natural. The tan was the only clue that revealed she wasn't the celebrity he thought she was. She bore a strong resemblance to Princess Gwendolyn, of the royal family of Great Britain, a daughter of King William, born when he was in his forties. Occasionally individuals from powerful families came to Annapolis, and this must be one of those times.

He had read about the princess who had taken the world by storm with her dark, wavy hair that featured red highlights. Her green eyes were wide set and a perfect nose set off her high cheekbones. "Gwens," as she was affectionately called by the press, who in a previous generation had called her father "Wills," had charmed Englanders and Anglophiles the world over with her royal elegance mixed with tomboy mischief. There was heavy speculation as to whom she would marry. Every month it was a rich playboy, movie star, soccer player or sixty-year-old magnate. Unlike another popular Princess from the last century, Diana, there was currently no prince in the picture.

Sam shook his head. He was being silly, of course. The young woman in the isometric booth sure did look a lot like Gwens, except this one was prettier. Sam was mesmerized by her body in slow motion as she twisted her torso and spread her arms outward. Her eyes were closed and her movements, graceful under the heavy gravity.

"Not bad," he said softly.

With Jake Addington and his tests temporarily forgotten, Sam got into his own isometric booth and felt the heavy, invisible waves push him down. With supreme effort, he raised himself on his tiptoes. He held his arms out. He pretended to concentrate on the exercise as he kept one eye on the beautiful princess look-a-like.

After about five minutes, a beep sounded and the woman opened her eyes. She stepped out of the booth and wiped her face and shoulders with a towel. She didn't notice Sam, who was now outright ogling her. With growing interest, he watched her exit towards the cinder path that wound around a large part of the academy grounds.

The isometric booth would not allow him to finish until it "knew" his workout was complete. It would not be finished until every muscle was worked and stretched. There was an abort button, and with an all-out effort, Sam took a step towards it. Sweat poured out of his body like a squeezed sponge as he strained forward. Movement caused the forces to increase and the muscles on his neck and arms rippled as his stiff fingers slowly made their way to the button. Finally, he pushed the round, gray switch.

A loud beep sounded and the incredible weight lifted off him. He hopped out of the booth and toweled down. Sam took a breather and let his muscles adjust to normal gravity. After a moment, he headed for the exit.

The princess was a blue dot in the distance. He could see a practiced, athletic stride that even he did not possess.

"I'm faster though," he said with determination. He took off down the cinder path. His legs felt heavy, having run a couple of miles through Colorado, resisting the isometric booth's force. Sam's adrenaline level picked up while keeping his eyes on the prize. She was definitely worth chasing.

It took four minutes for him to catch up. He admired the snug way her blue shorts fit her body. Nice butt, he thought. Despite his protesting lungs and hesitant legs, Sam picked up the pace and quickly passed her.

"Nice butt," he heard her say.

He smiled and let her come abreast of him.

"I hope you are not my superior officer," he said.

She kept her eyes on the path ahead. "We're all equal on the jogging path."

"I must admit, I had equal thoughts as I followed you. Did anyone ever tell you that..."

She cut him off. "Yes, Princess Gwens," she said with exasperation. "The woman haunts me."

When she looked at him, he noticed that her eyes were blue, not green like the princess.

"You're prettier."

"Thanks," she said, as if the phrase were hardly original.

It wasn't original, and this athletic beauty had put him in his place, Sam decided to jog quietly for the next hundred yards or so. Then he spoke. "That's what they say, huh?"

She nodded and laughed.

"Let's talk about something else. You have a nice stride. Pretty fast, aren't you."

"Glad you noticed. "I needed that," said Sam. "This hasn't been a very good day for me."

She smiled, even though her voice got serious. "Oh? Do tell."

"I found I have to take a retest for a mission assignment I'd set my heart on."

"Poor baby. At least you get a second shot. Sometimes when you washout, that's the end of it."

They jogged further and then she slowed and stopped. She leaned against a fence railing.

"Was it the Polaris project?" she asked.

He looked surprised. "Yeah, how did you know?"

She began walking ahead of him. "Everybody wants to be on that project. I mean, come on, colonizing Mars? It doesn't get any better than that."

He eyed her suspiciously. "Were you chosen?"

She looked like she was deciding on what to tell him.

"Yeah. I'm going to command the Sunrise, the supply shuttle for the Alpha team."

Sam shut his eyes. The Alpha Team was his assignment of choice. They would be the first to remain on Mars, laying down the ground-work for the colony.

"You're one of Addington's men."

She smiled broadly. "Excuse me? A man? "

"You know what I mean. That prize jerk probably hired his brother-in-law and college roommate to pilot the shuttle. Just give me a few minutes alone with Mr. Addington and I'll show him how the real world works. You get the best people possible, not a bunch of trained monkeys."

She gave him a serene smile.

"As a matter of fact, Commander Addington did train me. He hand-picked every man, woman, for the job. There are fifty women in my detail alone."

Sam started to jog again. The woman seemed less alluring. She was one of Addington's "men."

"Hey, wait up," she said as she kept pace. "I'm not going to defend Addington; he can do that himself. If you have a problem with him, I couldn't care less, but I don't like your insinuation."

"What insinuation?"

"That I'm part of Addington's good old boy and good old girl net-work. I earned my spot on the Alphas."

Sam waved her off.

"Don't mind me. I'm just disappointed. At least I get another shot. "

They jogged another two hundred yards before Sam spoke again. "I just don't like the guy. I think he has it in for me." He stopped once more. He had to take a breath.

"Are you tired?" she asked.

"No, I'm just thinking about collapsing here on the jogging path." He held out his hand.

She shook his hand.

"I'm Sam Mickelson," he offered.

She grinned. "I'm Eliza Addington," she said.

His eyes grew wide with surprise. "Oh," he whispered.

After that, Sam and Eliza spent most of their time together. She insisted that he have dinner with her father, which turned into a stiff affair. They had met before, but the senior officer wanted to keep any hint of favoritism out of their relationship, as he had done with Eliza. Others had tested her. He would personally test Sam for the second time.

But there were moments of supreme beauty, social events such as the Senior Graduation Ball where they turned heads as if they were movie stars. Even the band fell silent for a moment as the pair dressed in formal Naval attire walked out to the dance floor and waited as if their moment were choreographed. There was the conductor's downbeat, and the musicians started again, playing the brightest, most beautiful music they knew, an old Johnny Mercer tune called "Moon River," totally appropriate for the pair of high flying wanderers. Sam and Eliza danced as one while others fell in around them. It was a magical time.

Soon after, it was time to launch their careers. There were no wars going on at the moment, a few skirmishes in Sierra Leone, which was not new, in Venezuela and Saudi Arabia, but of course when you are in the military, you are always at the beck and call of your superiors, and you go where your orders send you.

They already knew that she was named to the new astronaut corps whose destination had not yet been announced, but everyone had guessed that it was Mars. The Aquarius mission using robotic equipment developed by energy companies for mapping geologic formations, had led the way to the discovery of large underground aquifers there. A Martian lander, the sixth one launched since 2015, had the ability to process soil for chemical composition. The space scientists believed that they could begin their colony, and they had been right. The work had gone swiftly and the station was populated by 2030. He expected to go where she did.

But something happened that altered their plans.

Once again, at the NASA training facility, he was put into an isolation capsule for twenty-four hours. Horns blared unexpectedly. Vapor was released. A variable hum backed by flashing lights was piped into the tight space. Finally he was subjected to a series of in-flight challenges that he had to solve, since failure was not an option. There were instrument outages, engine misfires, re-entry dilemmas and any other disaster that the director of the program, Commander Jake Addington, could devise. What a jerk he was, Sam thought. He told himself that it was because he was dating his beautiful daughter, Eliza. But he knew too that the challenges required the best minds, and those with the steady nerves of a diamond cutter.

It was the last eight hours. Sam had kept up with time as best he could, and he was sure he could make it through, and in his mind, be worthy of Eliza. Both of them in the Space Corps. Wouldn't that be something to talk about on date nights.

There was one memory that he held to tightly in order to maintain his sanity and complete the test. He had played quarterback for Navy, and he loved to remember those days. The roaring crowds. The pomp and circumstance. Taking the field with his buddies. He'd quarterbacked the Midshipmen to a bowl game in Dallas, something that hadn't happened that often. The final play was what he remembered. Gathered in the huddle for the last time, the score 13-7, the game appeared lost. Sam sent the wide receiver on a turn-in pattern. He turned to his O-line behemoth. "If you let that linebacker through one more time, I will personally kick your ass after chemistry class Monday. All right. Go!"

How could he lose?

After the game, the white hats went high into the air, and Sam owned an achievement that he would never lose.

The sheer force of his will kept him in the chamber until the last hour. His blood pressure shot up, and they pulled him out.

Addington was waiting for him, his military demeanor in place. "I'm sorry, Sam. This is not for you." He didn't wait for a reply but did an about-face and walked away.

Sam stared after him, clinching and unclinching his fists. He would not be a part of the Polaris program. The military had something else in mind. For him, whatever they decided, it was a consolation prize.

NASA had grown even larger in its financial base, finding support not only from the United States government, but also from other nations and most importantly, corporations who wanted to hook their own space futures to the United States' ever rising star.

The U.S. had decided to leave the moon alone. Already there was a station there, and many businesses and governments were using it, mainly to shuttle wealthy individuals to have the adventure of a lifetime. The Walter Cronkite Moon Base had become more popular than Europe as a vacation destination. Images of civilians bounding around on the moon in space regalia was broadcast across the planet, free advertising that was priceless.

Space technicians had long since solved another riddle, how to keep the space travelers from losing all muscle tone. They accomplished that with an electrical exerciser that added enough tension to workouts to maintain the muscles.

After he washed out and before Eliza headed for Polaris where her father would soon be in charge, they spent as many hours together as they could. The sting of not being selected for Polaris while his girlfriend was receded as Sam got to know Jake better. The older man had taken a leave and was spending it in Houston where he would soon be training young astronauts he had chosen first for Polaris and later for the Mars expedition. He was often at the Space Center where Sam awaited his orders.

Both men discovered they had a common love for the Dallas Cowboys. It was an enjoyable bonding experience. They both loved Annapolis, flying and shared an interest in the environment. Jake soon let Sam know he trusted him. Eliza had truly lost her heart to Sam.

Sam at last learned his destination, school in Norway, his assignment the study of cross hybrids taking the place of natural plants. A special laboratory had been set up to simulate conditions on earth as well as on Mars. The other part of the experiment was to measure how many hybrid seeds would be required to feed two hundred to three hundred people. The plan was to think ahead, in case of terrorist attack on the food chain.

The good part was that Sam would be in professional contact with Polaris, which was better than nothing.

He wanted the assignment, but he also knew that he would miss Eliza. They parted reluctantly, but they believed that their separation would not be permanent.

Once he was in place in his station in Norway, and she settled in Houston near the Space Center for further training, they could communicate with ease. Even though she loved her work, it was always fun to return to her quarters and find messages from Sam telling about his day, asking her about hers.

A year passed before they received furloughs and decided to meet in London which was a short trip for him, a little longer for her. He was twenty-four now, and she was twenty-three. He found himself wondering when they could be together permanently, what the 'success' they both desired would look like, and if she might consider deferring her career to his, the answer to that he knew already. She would not.

He was waiting for her at Heathrow International Airport near London on a Tuesday morning. He'd made that transatlantic flight before, and he knew how tiring it was, but when he saw her she was practically bouncing with joy at the sight of him. For him, seeing her caused him to draw in a deep breath of something like awe. She was as beautiful as ever, as vibrant as he had ever seen her, and he was more in love with her than he could remember.

He splurged on a taxi, to get back to his hotel more quickly rather than to show her the sights, but on the long ride in, he did that as well. He held her hand tightly and could not stop himself from staring at her. He could tell that she liked the way he seemed more attuned to her than to the grand sight of stately London in the morning. How beautiful it all was, fresh and new to both of them.

"You know, we have five days to see the sights of London. I can't wait...," he said.

She squeezed his hand. "Me too."

Neither was talking about the sights.

From a distance, Sam pointed out the Tower of London, the Tower Bridge, the ships cruising along the Thames, St. Paul's Cathedral where they would take communion on Sunday, Westminster Abbey, Winston Churchill's War Rooms from the 20th century, all the time holding tightly to her hand. Once when they were staring up at the Tower of London, a local shopkeeper came up to them and wanted to know if she were a sister to Princess Gwen, the daughter of the King of England. They both laughed and said, "Distant cousin."

At last they reached the hotel where he had already checked them in. Soon they reached their room and stood staring at each other while they waited for the luggage. He took a step toward her and took her in his arms and they were kissing when the boy arrived with the bags.

Sam grinned and released her but only for a moment. He opened the door and allowed the luggage to be deposited. He gave the young man a tip, and told him to skip the lecture about how the room operated. The two of them were electronic wizards and needed no such instructions. In a few minutes, they were alone.

Few words were needed. He began to undress her and she reciprocated. Both had waited a long time for this moment.

He would remember nothing and everything from that day. Together they lay face to face on the bed, and high above he could see the colors of the room, soft oranges and reds, against the drapes that he had pulled tightly closed. Everything was neat and clean. The place had been recommended to him by other service people, an inexpensive hotel that was well kept by a former British Army Colonel who remembered his own days as a young soldier. The walls were a golden color, a rough texture changing the way the light was reflected.

Of course there were communications devices, everything from an HD-3D television set to a hands-free phone system. Somehow he had hardly noticed that, thinking that their time could be better spent.

The outside sounds were muffled, taxies, buses, fire engines, air conditioners. A bright light was still on in the bathroom, and he turned away from that. The rays of the diminishing sun were coming from the window, through a slender opening sliver in the drapes.

None of that mattered. He carried her to the bed, and the afternoon faded into evening unnoticed. They had waited so long, but no more. She would ever be a part of him, and despite the incredible distances that would soon separate them again, she was his and he hers. All of their ambition and all of the success that would inevitably come, would fade in light of this love that had consumed them, and now bound them forever.

The week sailed past. Every walk they took was a revelation, Charles Dickens' home, the crown jewels kept in the ancient tower for hundreds of years, the Shakespearean play, Twelfth Night. They marched along Sherlock Holmes' stomping grounds declaring that the "game's afoot." They walked reverently through the great Halls of Justice, where they noted that some of the barristers still wore long white wigs. They made the trip up the Thames to see Hampton Court and down to see Greenwich, where Eliza stood proudly with one foot on each side of the Prime Meridian, encompassing both sides of sidereal time simultaneously.

Back in London the double-decked bus rides delighted them, allowing them to see a wider view of London. They had their photo made in an arcade and Eliza announced that it would be on their wedding invitation, which would be held in the chapel at Annapolis. After all of that they returned 'home' and made love.

"Did you mean that about marrying me?" Sam wanted to know.

Eliza smiled. "Of course. I planned it from the first time I saw you, nice butt that you have."

Too soon, the week was ended and it was time to return to the real world which they had chosen.

At the end, they said goodbye at the hotel. She was returning to the United States and continued Astronaut training as module pilot. He had his new assignment in Norway, at a second seed bank that was being developed there in a series of deep caverns not so far from the North Pole, near the first one. The older one had run out of space. And there were technologies the scientists wanted to try out in the new facility.

Eagerly, as young lovers will, they planned to meet in Dubai in the fall. They kissed so long that the taxi driver actually tapped his horn. She grinned at Sam, turned and got into the taxi. She handed him a small bag from a leather goods store. He looked inside and saw a pair of beautiful leather gloves. She waved, and blew kisses, and he mouthed, "Dubai."

It was the last time he would see her.

Twenty Nine

Expectations

In no time, they were both involved in their work again, he in Norway where he wrote that it was extremely cold up in the area of the international Seed Caverns. She wrote the opposite from southeastern Texas, that the city of Houston was so hot and humid that she had to wring out her socks every morning, a slight exaggeration but meant to make him laugh. Everyday they would mark the calendar. Dubai, here we come, one would write, and the other would repeat the phrase the following day. Still the work did not suffer, disciplined as they were.

Four months after they had seen each other in London, Eliza's greatest dream was about to come true. She was to pilot a supply ship to the new Polaris, her job to deliver the payload. There were plans for three such missions. Except for her time with Sam, she told him that she had never been happier.

The training period went by swiftly and Eliza had proved a good choice for the mission. The day came at last, as she was deciding that time was standing still. It wasn't.

There had been a short period of intense interest from the news media for her mission. A lot of reporters had drifted away to some other subject, since space travel and women pilots had become commonplace. Still, there were reporters from many nations, particularly South Africa which had produced two of the current astronauts. Truth was, there was still nothing more spectacular than the launch of the huge rocket ship blasting off its launch pad with a roar that could be heard for miles.

Few remembered the iconic newsman from the 20th century, Walter Cronkite, who had wanted to go into space himself, but was never allowed to do so. Wouldn't a man like that have loved seeing the group going now, including the ebullient Eliza.

The countdown began and everything proceeded as it usually did. Dozens of engineers studied their liquid computer screens, checking with each other, marking the time as they always did, until the final moments. Eliza could hear it all, the numbers, the roar of the engines, something ticking like a clock. With everything within her, she trusted the technology of the space program. It boasted the best technological minds from a dozen countries.

The one thing that bothered her was that she could not see much, the newly designed helmet fully protecting her head and eyes.

What she did see was a small screen inside the helmet and the numbers moving backward, eight...seven...six...five...four....three.... two....one....." And the roar of the space ship lifting off the pad and flying toward the sun. She had seen the takeoffs enough times that she could visualize the billowing cloud of dust and smoke and fire. They were on their way.

The crew was welcomed warmly at Polaris. It required two days to unload the gear, some of it to maintain the station, another part of it to go on to Mars once that mission was deemed ready. The smaller space ship that would carry the Mars crew on had to be assembled on Polaris. Since the Mars vehicle would not be subject to atmosphere, it was shaped differently from the Sunrise, which would return to earth for its next payload.

The days passed swiftly, and it was time to go back to earth to load and deliver more essential parts of the Mars effort.

THIRTY

An Unforeseen Event

Ketumil Luba was surrounded by the press. He had learned to deal with the constant attention and seeing his face all over the internet. The world had chronicled his every move since he left Botswana and entered the astronaut training program. He had been labeled "Africa's First Star Voyager."

He was from a family of shepherds in the Kalahari. Called "Mil" by his brothers, as a young boy his eyes were always on the heavens. They swore that he had the best vision in all of Africa. The vast desert sky was his planetarium. On a trip to Gaborone, the capital of Botswana, he walked into a bookstore and spent two months' wages on a volume of astronomy. He learned about the history of the early star gazers such as Ptolemy, Galileo, Tycho Brahe, Johannes Hevelius and Johan Elert Bode. He soon knew the names and locations of stars and constellations on any given day.

After memorizing the treasured volume, Mil's father rewarded him with two more books. The young shepherd boy became a human encyclopedia about astronomy. His brothers liked to ask him questions in the presence of an unsuspecting stranger.

"Mil, tell us something about space."

Without thinking, Mil spoke. "You can see Venus and Mercury with the naked eye. Every January 26th, Mercury is low in the east before sunrise and it pairs with Venus."

The brothers always smiled when they saw the stranger's perplexed look.

Mil caught the attention of the local paper and in a few weeks, the fourteen-year-old's story was all over the internet. He had been recruited to Cal Tech where he studied, what else, astrophysics. He was an apt student and now, Mil was a full-fledged astronaut.

Members of the media, particularly those assigned to the science beat, loved to question him. "Mil, do you think you were chosen for the Alpha Team because of your unique circumstances?" asked a reporter. Mil spoke four languages, including his native dialect, and English was his best.

"I am responsible for construction on Mars. My team will consist of fifty workers who have trained underwater, in vacuum environments, and isometric rooms. We will build a small city on Mars in less than a year."

The reporter was not satisfied with Mil's answer.

"Yes, but I'm sure Commander Addington has plenty of candidates who..."

Eliza stepped next to Mil. She wasn't going to let the words "more qualified" be uttered by the reporter.

"Mil has a degree in Astrophysics with an emphasis on space architecture. He designed a computer program that not only is creative, but practical when it comes to building the structures we need on Mars. When Mil joined the program, my father chose him because we need him. No one else has this type of training."

The reporters had already questioned Eliza concerning the possibility of nepotism created by Jake's naming her to the elite team. She had satisfied their cynicism throughout the phase 1 programs of Alpha team training. It was obvious that Eliza knew her job and did it well.

The space shuttle, "Sunrise," was the largest ever constructed. It was jokingly referred to as the "Titanic" among the team members.

"I just hope we don't end up like the Titanic," Eliza commented.

Joel Kempler, a pilot who would be making the round trip with Eliza, looked up at the two hundred-foot shuttle. "It's a lot of mass to deal with. We've come a long way since the last century."

Eliza nodded in agreement. "I've seen old videos of those first disastrous blastoffs. It was like putting a bullet on top of a powder keg. As much as I know about it, I am still amazed what we have accomplished."

They looked up at the Sunrise shuttle. It looked like a blue whale standing on its tail. It would carry fifty members of the Alpha Team. They were the leaders and supervisors of all departments. They would lead the first attempt to colonize Mars in every aspect. Sunrise would make two more trips back to earth to pick up the workers and support personnel, in addition to tons of raw materials, computers and food.

"Are you excited, Joel?"

"This is massive. I've trained for this all my life. Like you, Captain. I've had my eyes set on tomorrow. Just think. They'll be creating a whole new world up there. Someday, if there is some disaster endangering earth…"

"Like a giant meteorite…"

"Exactly. We bring people to Polaris and then on to Mars."

It was unstated that this was the ultimate goal. The immediate goal was to grow food, develop a water purification system and create a life sustainable environment on a hostile planet.

"Get some sleep, Joel Kempler. Tomorrow, glory awaits."

He gave her an affectionate hug and they left the press briefing where the reporters were interviewing Alana Gibson, chief medical officer and unofficially, the first surgeon general of Mars.

Eliza rested in her NASA quarters going over her checklist. In the old days, she and her crew would have been isolated before going into space. Inoculations now took care of that, making all interplanetary travelers invulnerable to earth's common diseases. Polaris residents were amazingly healthy. Of course, most of them were young and always in training.

She spoke to her father on her videophone about the mission. His face was slightly distorted from the lens, but he looked good. His voice came through the walls clear as crystal.

"Hey, honey, it's good hearing from you,"

"Hi, dad. I hope you've got a reservation for us at The Landing."

"I thought you were bringing lobster up here with you."

"We are. And enough meat to keep all those fast food chains up there busy."

"I've got your arrival at 12:43 a.m. earth time."

"We're on schedule."

"Have you talked to Sam?"

"Last night. It's weird. I can talk to you up there, but the transmission is fuzzy down in Antarctica."

"Good old rock and distance can't be penetrated sometimes."

"Mil is confident we can get started in two months."

"Actual construction? We'll see."

"I've got most of the materials in the Solar Section. Three more trips for the Sunrise, and everything should be available. Mil and Capt. Lewis handed the sheets to me yesterday."

He frowned, which looked worse over the transmission. "Well, check again. Check it yourself tonight."

She saluted her phone. "Yes, sir."

He looked at her for a long moment. "You look exactly like your mother."

"Oh? Did she salute you often?"

He laughed. "More like the other way around."

Eliza did not know her mother, other than the pictures and videos of her. Captain Charla Addington had outranked her husband. She was an astronaut like her daughter, but ironically she had died in a civilian air crash. Eliza was two months old. Now she was a couple of years older than her mother when she had died.

"I'll see you on Polaris, Dad. Gotta get some sleep."

"Goodnight, pumpkin."

"That's Captain Pumpkin to you."

Just as she was about to switch off, Jake spoke. "I'll be at dock seven on the upper level. You'll be coming in slow. Look for me. I'll be wearing my red service overalls and blue cap."

"Okay, I'll be in the forward chair with Joel and Chip Henderson. First window, look for me." First window was wide and he would be able to see her.

"See ya soon, Dad."

The Sunrise experienced no difficulty at lift off. All systems were go and when the huge ship escaped earth's gravity, Eliza felt like she was finally on her way.

It had been two hard years to train for the next two years. Getting engaged to Sam made it all that more special. They had little time together after he left for Antarctica. He was there on a four-year assignment, but the plan was simple. They would get married in twelve months and her father would have his new son-in-law transferred to the Mars unit. The work he was doing now was very similar to what would be required on the Red Planet, although of course there was not yet any agriculture on the Red Planet. It was all perfect. Aboard ship, she received a hoped-for call. She saw Sam's face on the screen.

"Pick up, Princess."

She answered. "Hey, you."

"Looks like our reception is better when you're in space."

"I'm missing you already."

"Back at you," he said.

They just stared at each other on their phones, content in their silence. Then she spoke.

"How's everything at Ice Station Zebra?" That had become their pet name for his South Pole station. Officially it was called Vinson Base, because it was located near Vinson Massif, the lowest elevation on the continent. Sometimes Sam referred to it as his "Fortress of Solitude."

"Are you wearing your woolens?"

"Are you nuts? It's always hot here. I think they're trying to compensate for the temperature outside."

"Are you having any luck?"

He looked back up at the screen. "Corn seed."

"What?"

"I've got a seed corn under my microscope right now. It's got some unnatural properties."

Eliza was perplexed.

"What do you mean?"

"You remember your history, the Irish Potato Famine in the 1840s?"

"I've read about it. Fill me in."

"Irish farmers depended more and more on a potato that had traveled into Europe from the Andes. Two hundred years before the famine, Spanish explorers brought their potatoes and these varieties possessed little or no resistance to the diseases that were their enemies." She finished for him.

"So the potatoes rotted in the fields which caused a mass exodus from Ireland, but not before two million people starved to death."

She could see the look of pride on his face.

"Very good, doctor. You score an A." He continued. "I met a man during my undergraduate studies. Dr. William Fry of Cornell. He read an article I wrote in the Maryland Ag News about experimenting with a chemical that gave a virus to antibodies that cause a tuber cancer."

She smiled. "That was a brilliant analysis."

"You read that?"

"When we first met, I went back and read everything you ever wrote." He looked pleased, then continued his story.

"Dr. Fry invited me to a farm in Potter County, Pennsylvania. He showed me a field he had been cultivating. At first, I didn't know what I was looking at." Sam's face turned distant as he remembered that day in the field with Dr. Fry.

———

Dr. Fry, a small wiry man who looked more like a garage mechanic than a world class scientist, pointed to the field.

"Do you see it, Sam?"

Sam scanned the rows slowly. Then it hit him.

"They're dying."

"Dead, Sam, dead. Only days ago, this field was a lush, green canopy of potato plants, but almost overnight, they turned into rows and rows of rotting leaves and gnarled vines."

"They look like they've been torched, sir." Fry nodded.

"Yes. This field and all two hundred acres beyond that."

"What happened?" Sam asked.

"Our culprit was a strain of Phytophthorn infestans. It's a version of the destructive organism that caused the Irish Potato Famine."

Sam gave a low, impressed whistle. "Incredible, sir."

Fry continued. "Except this strain took a biological leap. It's a new fungus. If it's not stopped, we could see a famine of epic proportions in the future."

"How did it happen?"

"The new fungi spread from some 'new food' potatoes over in the next county."

Eliza was looking at Sam's face as he related his meeting with Dr. Fry. "What does that have to do with the corn seed?"

"Well...sometime in the last century...the 1970s, a fungus called "bipolaris maydis" hit the corn industry. The "Southern leaf blight" overpowered millions of acres of corn and in one year, wiped out nearly twenty percent of the nation's corn crop."

"I never heard of that."

He shrugged. "It was before our time."

"But how could it happen?"

"The uniformity of the crops by then was the enemy, leaving a subdued scientific community in its wake along with a dearth of corn products for several years."

Eliza saw where he was going. "The key lesson was genetic uniformity is the basis of vulnerability of epidemics."

He was quiet for a moment. "You get another A, Princess."

"So Sam, this corn seed you've got reveals some similar properties. You're working with all sorts of hybrids. It's expected. Do you think a new strain has developed in this sample you've got now?"

"Eliza, this sample is from the natural bin."

She felt a chill. "Do you think someone tainted the supply?" She saw him shake his head.

"I don't know."

They had to end the conversation. Nothing was decided, except that she encouraged him to find allies at his base.

THIRTY ONE

Star Bright

Polaris looked bigger in the window. Several modules swarmed slowly around the port side as they got nearer. Eliza joined Joel Kempler and Chip Henderson at the controls.

"How are we doing, Captain?" Joel gave her a broad smile.

"ETA is ten minutes."

A smaller shuttle took off from the port where they would be landing. Space travel was so quiet, Eliza marveled to herself. Getting close, she could see three bullet trains coming out of their tubes and speeding along a huge boulevard. It had to be at least a hundred yards wide. But there was a soft silence. If they were on earth, Polaris would be as loud as Tokyo's Ginza or New York or Beijing.

Now, the entire program had been set in motion and Eliza was part of it. "How are we doing, Captain?"

Joel gave her a broad smile. "Five minutes."

———

Jake had paced up and down in his office, unexpectedly anxious to see his daughter. He stopped at Out of This World gift shop and bought a charm made of moon rocks. In the old days, he felt bad when he left his little daughter with a caretaker while he traveled the world, doing his duty. He started the tradition of bringing her home toys and plastic knick-knacks of little value. Even as a teenager, Eliza looked forward to these treats, despite their silliness.

Jake put his hand on the scanner and his account was charged for the charm.

He put the bagged bracelet on the seat next to him and headed for the upper level of the port.

The Alpha Team would exit to the sub level, but Jake wanted Eliza to see him. He wore his red service overalls and blue cap. When the shuttle docked, he would have time to go below and greet his daughter. He had seen her six months before at NASA. He remembered how nice it felt to have the earth beneath his feet. He was conducting an orientation class for the trainees. His last night, he had dinner with Eliza at The Breakers in Cape Canaveral. He had a little difficulty getting her on point since she was expounding on how great Sam was doing down in Antarctica.

———•———

Eliza had her hand on Joel's shoulder as he sat in the pilot's seat. Chip Henderson was on the line talking to the port navigator.

"Thirty degrees by seven-o-three. The angle looks good."

"Continue program, Sunrise, and welcome to Polaris," came the voice over the speaker. Eliza could barely contain her excitement.

"We're about to begin a new chapter in history, gentlemen."

"Roger that," said Joel.

Mil came up from the lower deck.

"I just saw my homeland and my spirit guide beckoned me from the desert. I saw him."

Eliza smiled and shrugged.

"We'll find a little corner on Mars that we'll call "New Botswana.""

She walked over to the window and looked out. She was not surprised to see a red clad figure in blue cap waving his arms.

She smiled and waved back. "Dad!"

He stopped and took something out of his pocket and held it up. It looked like a necklace but it was too far to tell really. She felt a lump in her throat and waved again. He threw her a kiss.

Eliza did not find it awkward to interact with her father, even though he was head of the station. They had two delightful days. She told him about Sam, and their plans to marry. Jake told her that he would join her wherever they elected to have the ceremony. She hadn't decided yet.

At dinner their last night, he reached across the table and lifted her chin.

"It's been so long since we've been able to be father and daughter."

"You're a taskmaster, but then again, that's why the Mars expedition is going to work."

The time passed swiftly, and then it was time to return to earth. Eliza finished preparations, loaded her crew aboard, and ran through her flight plan. The Mars voyagers were staying on Polaris, of course.

Aboard the craft, Eliza was pleased with the relatively gentle launch from the station. She decided to have the craft perform the expected backward flip which appeared to be in slow motion. It was a maneuver begun many years before that had become a solid tradition.

Jake Addington had left the command center. It was just a short golf cart ride to the terminal. He saw the Sunrise doing its graceful backward flip, and wanted to be in view so Eliza would see him if she could.

She could see her father below, waving at her. The Sunrise moved away from the station. Nothing seemed amiss. Joel was concentrating on the instruments, but something caught his eye and he spotted the piece of space junk the size of a Buick automobile hurtling toward them on an unavoidable collision course. He tried to warn Eliza, but when she saw him turn toward her, a look of alarm on his face, she looked puzzled. She glanced over the monitor, the dials, but it was too late. The shuttle went white inside.

At first, Jake thought it was a power surge. Then he realized what he was seeing, what first appeared to be elaborate pyrotechnics, but it was not that. The explosion filled the sky with streaks of burning light.

When Jake realized what had happened, he began to weep. He knew that he could not even retrieve his daughter's body.

The public, accustomed to the hype from television news and Hollywood tales that always had a happy ending, expected perfection. They didn't want to know how complicated and dangerous the work was.

It would never be explained, despite all kinds of suppositions, but it had occurred in full view of the world. Such a scenario had always been a possibility, had happened before, to another of the new shuttles when they were first launched, astronauts trapped, and finally doomed.

What greater negative impact could the destruction of the nation's pride and joy have than to be seen around the globe on CNN Intergalactic Television.

Sam learned of Eliza's death that same night. After months of inquests, committees and second guessing, it was declared an accident. Mil wept for days.

PART II

Thirty Two

The Strongest Man In The World

Angel had started searching for others who were still alive near his home, and he was surprised at who he found in the county.

First, there was Chuck Ames, the football coach, a great strapping man with a love for the game of football. When he began to lose his players – and some disintegrated quickly – he set them free from any obligation to play. When his entire family died before him, he moved his things to the high school field house and camped out there. It was where Angel found him. They agreed to help each other, and the first thing was to find out if there were others. They began to call themselves the "Old Ones," as if giving themselves a name might stave off the horror that was destroying them.

Together Chuck and Angel rode the back roads of the county, gravel byways little more than lanes, searching for others as strong as they were. They first discovered David Sloan, a former professional quarterback who had spent his later years promoting healthful eating and exercise. Apparently they had worked for him, because while his hair was white when they discovered him at the ranch and retreat center, he remained physically strong. They recruited him to join them "Might as well. Looks like everything is falling apart," he said.

The last active member of the surviving group was a nurse, Wanda Owenby, who had migrated from Dallas to Rice more than twenty-five years earlier seeking a simpler existence.

Angel knew her from working at her farm, a small 10-acre place that had once featured the most beautiful strawberries and peaches in the state. He had often admired her high energy, the fact that she held down a job, gardened like a professional, and managed to hand out fruit to all her neighbors, and to the recipients of Meals on Wheels in the community.

After the virus had swept through the town, she had remained at the local health center which she attempted to keep open. She had come up with palliative treatments, vitamins, aspirin, banana shakes, innocuous things that did no good, but gave the sufferers a few moments of respite. But now, she had no more clients. Still she went to the clinic, hoping that someone would show up. She welcomed Angel when he came.

When he arrived at the health clinic searching for her, he noticed at once that she was giving to her right hip, and by that he knew her days were numbered. Still, she could help them for whatever time she had. She knew of one other, had visited him at his farm, but he was already bedridden. All they could do for him was provide something to eat, to get him through his last days. And Wanda still possessed medicines that would ease his pain.

Finally they knew that there were six left, Angel, David Sloan, Sam Mickelson who had only recently returned home, but he appeared to joined forces with the Ryder girl; Chuck Ames, the football coach, Wanda and the old man at the Mickelson place, who was bedfast. There was the young one as well but he was wandering around like a lost puppy and pretty much useless. He was still the strongest among the survivors.

When they went out to see the old man he knew he was dying, but he seemed determined to contact Sam Mickelson. It seemed important to him, although Angel argued that it was a waste of time. Still, Angel promised to help the old farmer.

Angel thus far had survived everything, but many in the town and in neighboring towns were dead, which meant there was no one to call, no relatives or friends out there. It was an incredible loneliness that was almost unbearable.

What had surprised the older people was the appearance of some of the youngsters. At first, they appeared untouched by what had devastated everyone else. Wanda for one had thought of separating them from those who were infected. She could not have known that such an action was hopeless.

For a time after they discovered each other, they made an effort to get to the basis of the blight. They went to the home of the scientists on Clay Street.

Thirty Three

Kill The Messenger

Everything Dr. Ryder had feared had come true. Once the virus took over the lives of almost every living creature, Dr. Ryder struggled to find some sort of cure, some way to stop the monster from spreading, but he found nothing. Even after he attempted to collaborate with others all over the earth, nothing helped. The only hope was that the scourge would run its course, like ebola or other of the other viruses that threatened mankind.

On a rainy Saturday morning, Dr. Ryder and Linda were cleaning the lab, when they heard someone outside. Dr. Ryder opened the door and frowned

Angel stood there, and with him a half dozen others, some of whom Maurice knew, and some he did not.

'Come out of there," Angel ordered, "or we come in."

"What do you want?" Maurice asked. He could see the gun tucked into Angel's belt, and the tall man next to Angel held a shotgun.

Dave stepped up. "You're a scientist. You brought this thing with you, didn't you, experimenting on our crops, our food source."

"No, I didn't. I was sent here to find out what was already happening, to look for an antidote."

Dave looked down at his feet and shook his head. He wasn't buying it, not a word, even though it made sense. He was looking for someone to punish for the error.

Chuck moved toward Dr. Ryder. Linda was standing behind him.

"Doc, come out and bring the missus," he said.

Dr. Ryder shook his head "no" and closed the lab door, hoping the men would leave. Instead, Chuck stepped up, pushed hard against the wooden door, and it fell off its hinges. He led a half-dozen angry men inside. The Ryders were unarmed and had no chance.

There was a loud boom, another second shot exploding, and then silence. Gun smoke hung in the air. Angel stared down at the pair of scientists. There was no question that they were dead. He hadn't meant to go that far, but what difference did it make? They were all going to die soon anyway. Dave wanted to bury them, but Angel walked away. It was another stupid mistake. The scientists had been their only hope.

Thirty Four

Jake Visits The Future

Jake had slept little in the past three months. Waking and sleeping he could think of nothing but his horticultural projects. He couldn't make the seeds grow faster, but he could enmesh himself in data, start new plant growth, and continue helping Colby from the great distance that lay between them. His old teachers, many of them military figures, always noted above all that Jake was the most determined man they had ever met. Now, working with little rest, he knew that it was the one characteristic that might help stave off the terrible course the planet seemed to be taking. He would never stop trying.

The last thing he did every night was head for the viewing room and stare at the earth. Each day, the spinning orb was changing, more parts of it turning gradually a rusty brown. The oceans seemed to hold fast against the pollution, their myriad blues and aquas less vibrant, but still clear. The lakes and rivers were mixed. Some seemed to flow cleanly, while others changed to drab shades of gray and brown.

It felt colder on the Polaris, although he knew it was his imagination. Nothing changed up there, particularly the heat and cold of the interior.

He could not help but remember the early work in plant genomes, when headlines appeared almost weekly about new families of genes that would fight off plant diseases. The hope was that the entire world would be fed. Even livestock breeds that were facing extinction were restored, and always, the genes and cloning played their part. No one suspected how overly optimistic the scientists were.

The same upbeat attitude prevailed throughout the agricultural world, from the rice farmers in the Arkansas and Mississippi Delta, to the dairy farmers in Decorah, Iowa, from the Texas King Ranch breeders with their Santa Getrudis and fabulous Wimpy P One quarter horse lines. Wherever animals were bred, no one considered that they might disappear.

The Jacob sheep with its lean meat, and its four horns, had been threatened in the 1980s, because it wasn't the most efficient breed, but some scientists had thought it worth saving, and did not buy into the popular notion that only the best of the breeds should continue along the ever changing path nature intended.

The truth was that all breeds were needed, as well as all kinds of seed. In the first years of the twenty-first century, once again, it became popular to depend on single breeds. The cow that gave the best milk. The sheep that provided the softest wool and the best tasting lamb chops. The chicken that laid the most and the longest lasting eggs. The most prolific seed corn. Don't worry, said the government experts. Everything is fine. Our work is progressive and above all, productive.

Soon, the unpopular fear of what could happen because of those decisions became earth's reality. The old adage to save the genes that were on the earth because your grandchildren might need them had come true. Only now, it might be too late. Native seed had been tossed aside, of no consequence. The horrifying thing was that it all happened so quickly.

Jake researched everything. He was surprised to find an article from 1993, in the sophisticated New Yorker magazine no less, written about hothouse tomatoes. A crazy thing had occurred. Americans, and that nation only, decided they must have tomatoes year around, and to get them, agronomists had devised a tomato that looked beautiful, but had a tough new exterior. The interior was all right, but often tasted flat. The exterior was almost indestructible and couldn't possibly be eaten. These hothouse tomatoes were available all of the time, and for a few years,

Americans settled for color and sufficiency. They stopped demanding taste. Some even forgot what that meant, but that wouldn't last.

The thing was, the new tomatoes were genetically altered. They became so common that when really good homegrown tomatoes hit the market in the correct season, consumers hardly noticed anymore. Was it any surprise that corn, peas, peppers and potatoes soon followed? At no time of the year was there a shortage. The companies that provided the vegetables earned four billion dollars a year. No one had any intention of leaving the path the industry had forged, that of altering food. A breakthrough product appeared on the market, and again impressed the consumers. It was called the "Flavr Savr," and it addressed the constant complaint that the "new" tomatoes lacked taste. Company officials said that they had integrated the "backyard flavor" back into the tomato. Their shareholders were excited, and they had reason to be. The stock of the small company that created this tomato soared.

Jake attempted to transmit everything he'd discovered and was still learning to Colby. Sometimes the transmissions worked. Sometimes they didn't. Would he ever see her again? On good days, he hoped so, but somehow, he did not think so. He couldn't help but wonder if Sam Mickelson was her friend or something beyond that, a relationship that he refused to consider.

Thirty Five

The Ring From Yesterday

Late on an overcast fall evening Colby sat beside the low burning fire, her head down, staring at the flames.

For a time, Sam watched her. Funny, a guy like him, known for his boisterousness, his love of life, could do so little to lift their spirits. He'd almost given up on that idea, but finally he spoke. "I found something I thought you'd like."

Colby showed little interest.

Sam fumbled in his pocket and pulled out a ring. He held it near the fire, and read the initials. "MR to LC?" He looked to her for an answer, shrugging his shoulders, clearly puzzled.

But then, Colby's face reflected her sudden engagement. She reached out her hand and slowly, lovingly took the ring. The slightest smile touched her lips.

"Yes. Maurice Ryder to Linda Cook. My parents. This is an old ring. My great uncle bought it for his bride, but she died and he gave it to my grandfather, his brother. It's been passed down ever since."

"Somehow your folks don't quite fit this little country town."

Colby was almost offhand in her comment, detached, as if remembering would be too hard. "I lived here my senior year. We were pretty much outsiders. Mom and Dad were scientists. Dad was the first to realize something was wrong with the hybrids. He saw the magnitude of the coming devastation."

Sam seemed to reflect her mood, disengaged, as if they were speaking of two other people. "But, clearly he couldn't figure a way to stop it. I know the feeling."

Colby's mood changed then. She was searching for something that eluded her. "Maybe he did. He'd sent me some ideas, some things to try. We were doing experiments on Polaris."

Sam looked interested. "And?"

"We got reversals, in the aging process. DNA replacement. The muscles in the mice regenerated. Eyesight. Hearing. Muscles regenerated. But it has to be tried out here."

Sam said, "On earth."

"Right," she said.

"If you were close to a cure and your parents knew it, why were they murdered?"

Colby flinched at the remark. The idea of the murder was hard to get used to. "Maybe...because they were outsiders? Or when they tried to give people hope, they didn't believe them. You know, kill the messenger."

"Why did you come back before you had a cure? Put yourself in danger?"

Colby looked at the man. She had to trust him. There was no one else. "My parents sent for me. They had found something else."

Sam leaned forward, clearly interested. "Worth coming back here for?"

Colby took a deep breath. "They called it the 'life source.'"

In one second, like a Texas norther hitting a balmy autumn day, the emotional climate changed. Sam was on his feet, his face livid.

Colby shrank back. She had not seen this before.

He paced back and forth, too close to her. "The best scientific minds in the world hadn't found a cure and you expect me to believe someone in these boondocks could?"

She began to hedge. She didn't like his anger. "Maybe."

Now he began to accuse her. "You could be snug on the Polaris, but you're risking your life for something..."

She got to her feet. Her voice was calmer than she felt. "An answer."

He moved closer. His eyes were hard. "Tell me? What is it? Where is it?"

She took a step backward. "I don't know yet! We have to find it!"

Sam's voice changed again. This time he was sarcastic.

"Am I to understand that we are on a treasure hunt? And if your brilliant father didn't make the directions too tough, we might get to live a while longer!"

Colby's voice was soft as late evening. "Yes." She heard the bitterness then. She hadn't expected it anymore than she had the sudden irrational anger.

"I was in the Antarctic with the finest minds, the latest information, the best computers and we failed! We even changed our mission in the search."

"To what?"

"Before, we measured how fast the ice moved down the mountain. We'd plant flags, figured it would take 100 years for the ice to move down the mountain to the coast...Global warming was the scare concept," he said. "Incredibly our nation and others joined together to stop it, to slow the glaciers movement almost to normal. Our flags had been moving too fast, from the highlands to coastal waters. We changed as inhabitants of the planet to try and change that and we succeeded to a point. But then, this came along. I suppose you could see it from space."

Thirty Six

A Lofty View

She joined him now, as if by their telling the story together, it would unite them. "This was different. Nature speeded up. We saw it from up there. Didn't believe it. Didn't want to believe it. All those predictions from the experts didn't cover this," she said.

His eyes stared away, at something in memory. "We saw the changes too. Usually glacier time is forever; it'd take three or four lifetimes to see any change, but what was happening was like a glacier nano-second."

She was remembering too. "Things changed ten times faster than the usual seasons turning. There was a period of incredible cold."

Sam segued into her story easily, as if they had the same memories.

"When we knew it was hopeless, there was nothing else to do about the virus, some of our people walked into the endless night," he said.

She frowned, studying him as if the answer to her question was in his face. "You ...didn't. Why?"

"I decided to come home...to die," Sam said.

She fixed her gaze on him. "Maybe you came back to live."

He shrugged, spoke in a flat tone. "Face reality."

"I'm doing that." She hesitated. "Trying to," she added lamely. She could see that he wanted to say more, but he fell silent. She watched him, felt confused and for the first time, a shiver of genuine fear coursed through her veins.

Thirty Seven

Team

She shook her head. There was nothing to say except, "I'm glad you had the guts to get here. No question about it, I need you." Her voice was soft.

He studied her face and thought, I need you too, and then he stood up. "Later," he said. He knew that soon he could tell her about Eliza. But now he wanted to be alone to think things through. He had struggled against nature to get this far and he was still standing. He did not want to die, and clearly Colby didn't either.

Thirty Eight

Learn From History Or Die

Sam remembered how it had been at the university. The scientists' hubris allowed them to forget what had come before, except their own myopic experimentation.

The story of the uniform corn crops was that for a time it was considered a miracle, that the hybrid corn allowed farmers to produce incredibly efficiently, but it opened the corn to the fungus which seemed to wait and then thrive on cool wet conditions, a perfect description of the summer of 1970. A lesson was learned just in time. The U.S. Department of Agriculture, the Food and Drug Administration and the National Academy of Science let everyone know what they had already experienced.

"The key lesson," both said, "was that genetic uniformity is the basis of vulnerability to epidemics." The word was spread. And the movers and shakers of genetic engineering promptly forgot.

At least some scientists were paying attention though, and the Fort Collins Seed Bank and other collections of native seeds were initiated. Still, the warnings had been there all along manifesting themselves in other animals and plants.

There was a question about certain butterflies dying, particularly the black swallowtail.

Researchers at the University of Illinois studied pollen from corn that had been genetically engineered to produce its own pesticide, a toxin naturally found in the bacterium Bacillus thuringlensis, or Bt. It was designed to kill a pest called the European corn borer which it did, but it also killed the larvae of the black swallowtail. The more important question was whether it would kill the caterpillar of the monarch butterfly. The scientists said it would not, and yet another warning was dismissed. One year Sam was on furlough and saw a small clipping in a local newspaper.

FORT WORTH, TX--Food Lion and Kash N' Karry grocery chains pulled their store brand white corn tortilla chips from shelves in response to concerns raised by the Food and Drug Administration about StarLink genetically modified corn used in the chips. The recall marks the first time the FDA has identified white corn products as containing StarLink. Detection of the yellow variety of the corn prompted widespread recalls last fall.

Sam had more than a premonition early on. Something bad was coming. Something greater than any war or plague or tsunami. Could anyone be safe from it? He had no answer for that.

The next event followed on the heels of the recall of the corn products. The Kellogg Company, whose huge corporate campus was in Battle Creek, Michigan, shut down its production lines in an unnamed location. The company's spokeswoman said that the company, proud producer of the popular Special K and Frosted Flakes cereals among others refused to announce any plans for reopening the anonymous plant, not wishing to give away the information for "reasons of competitiveness." The media had the story in its teeth and would not let go for weeks. The cause of the disruption, according to the Wall Street Journal and CNN, was that the giant grain elevators were unable to certify that their grain, actually corn, was not adulterated with the modified corn known as StarLink which had been bioengineered throughout the country in violation of what few federal regulations there were.

The Environmental Protection Agency began to issue statements but not warnings. "What we are hearing is a significant degree of concern about whether mills or food processors are able to provide a guarantee of no contamination or commingling with StarLink," a senior official said, and then added, "Because those guarantees are not being given, some corn is not being sold." On the other hand, a great deal of the hybrid corn was already in the marketplace. And what was contained in the corn was toxic to pests, and possibly to humans, but no one was saying that because the direct developers of the corn offered to buy back the nine million bushels on the market. It was far too late at that point.

Thirty Nine

"Heirlooms"

Knowing Sam's real story helped her although she sensed that there was more to it than he had shared with her thus far. She had not liked the idea that he was shallow and lucky, which had been the first impression of him. He was hardly that. He had tried to save his friends, but the task proved impossible. Certainly no one else had risen to the occasion.

There had been a publicized effort to get to the South Pole station via a specially equipped airplane, but that had failed. The weather had been too much for it. They were caught in a whiteout, and the ice on the wings pulled the nose down causing the airplane to slam into a mountain ridge when it strayed off course.

Everything planned for the South Pole station was destroyed, or at least gradually covered over by the ice and snow, hiding everything in the all consuming whiteness. She knew of it before she knew of Sam, and Polaris residents had been devastated when the plane went down en route.

Now, as the days wore on, Colby felt an anxiety she had not experienced for many years, not since leaving the familiar Virginia suburb of Falls Church and the beloved private school that catered to the children of the intellectual elite of Washington, D.C., youngsters like herself. When her family was assigned to rural Texas, it was a shock, and she did not quickly get over it. Her parents were excited to be going to the Lone Star State. They had heard nice things about Austin, which was the state capital, that it was a laidback and intellectual city because of the University there. Advanced research was being carried on there and at the neighboring Texas A& M University. But Austin was far from Rice, Texas, they were to learn quickly.

Colby also did not stay long in Texas, going away to school as soon as she could convince her parents to arrange it. It took only a few encounters with some of the young people in Rice, smart enough kids, a majority of them planning to go to college, but raised with a suspicion of anything or anyone who was different and Colby was that in spades. She wasn't interested in their music that much. It was mainly country and western, stories about lost loves, pickups and drinking. After awhile, she found it depressing and tedious.

She had been trained to play the cello, and there wasn't much call for a cellist in a high school marching band, although she had joined a Youth Orchestra in Dallas, the huge metroplex north a few miles. She had enjoyed that time with other classical musicians of all ages. There was a small Episcopal church in nearby Corsicana, and she occasionally played there, on Christmas Eve and a few other high Holy days.

But now, she had nothing to think about except waiting for the drop from the Polaris freight haulers, little to occupy her mind. On one early spring day, she came upon her father's hothouse. It was empty, clean, and inside were unopened sacks of potting soil, marked Organic. And with her, tucked away in a hidden pocket of her jacket were three packets of Burpee seeds, marked the same. They were not hybrids, and they were part of an experiment in space. Why not try them out? What did she have to lose?

There was another aspect in her thinking as she attempted to grow tomatoes that were not dangerous. Her grandfather, Mayor Ivan Ryder, had been famous in Falls Church for growing heirloom tomatoes the seeds of which his mother had brought from County Cork, Ireland, in the late 1830s. Her mother had given her the seeds after the family had eaten the tomatoes aboard ship. Colby's great-grandmother had been a child put on the Orphan Train, which took abandoned children, their parents dead or destitute, and delivered them across the country to families who would take them, usually for the work they could do. But the tomato seeds Colby's grandmother had brought were her secret.

Her name was Martha, and a family named Newlin had adopted her. They had an older son who for years treated her as a little sister, but that all changed, and one day they married, with the blessings of his parents. Martha had become dear to them, and they wanted to keep her in the family. She would often tell her eight children about crossing the Atlantic on a ship, and how her mother had died of severe dysentery and later, her father had had a heart attack while hauling barrels of beer at the brewery where he worked. That had left her all alone with no brothers and sisters, and the city of New York had placed her on the Orphan Train, going no one knew where. But she had landed eventually in Virginia, and become a Newlin for life.

When Martha had learned of Colby's interest in agronomy, she had planned to give the heirloom seeds to her. She had them specially packaged, but her time ran out before she could give them to her granddaughter. Still, when Colby had gone into space, her mother had given her the grandmother's gift. Colby had no idea then how valuable the bequest would be. In fact, it was a legacy of untold value.

At first she went alone to the greenhouse and sorted out the pots and seeds, labeling each thing she planted, making sure that it had the soil and light and water it needed. She prayed that the water would not carry the virus, but she would not know for many days.

There was something satisfying about digging in the dirt, planting the seeds, watering the plants, making sure they were in the right place for the sun. It all seemed a simple pleasure, and it was hard to think how it had slipped away from the residents of the planet.

She knew part of that story. An insatiable demand for food of all kinds, particularly fresh things, even when they were out of season. That meant that tomatoes first, and then peppers, squashes, potatoes and so on had to be grown to meet the demands of a six billion-plus population on the earth. The end game was all too certain. Growers forgot about the seasons of the earth, and grew plants and animals in essentially isolated situations. In 2010, scientists figured out the DNA of wheat, which had been a difficult and lengthy task, but the discovery opened the door to more experimentation. By the 2020s, 95 percent of the natural seeds had disappeared, except those held dear by the seed savers.

An odd story had appeared in the Associated Press. Mother Russia had what was the oldest seed bank in existence, dating from 1926. It possessed seeds no one else in the world had, but in 2010, it told the Seed Bank that its land was being given over to a housing developer and that it was evicted. There was a fight, and finally the seed bank prevailed. Colby wandered if it was still there. If it was, the Russian people had hope.

There were voices of reason, Greenpeace, the Sierra Club, Go Green troops and hundreds of campus organizations who quite literally could see the future. There was a Union of Concerned Scientists who begged the government to learn more before it unleashed the new foods, but they were not heard. Someone even mentioned a weed called 'kudzu' which had practically smothered the Southeast until a clever scientist from the University of Georgia had figured out a way to stop it. But it had taken years of research, and much of the land had been lost.

Certainly the Department of Agriculture was trying hard to keep up with the changes in food, but they were too few and had too little funding and authority. Despite the pleas for more monitoring, the development of the lab-created plants continued full–bore. The tragedy from the lack of vision on the part of government and agribusiness was that not only natural seeds were lost, but so were endless traditions surrounding food. Always, food played a huge role in understanding, even preserving, the traditions of cultures and ethnic groups.

In years past, the Smithsonian Institution had urged the preservation of the past by saving recipes which if they were old enough were called 'receipts.' The venerable institution searched the land for authentic cookbooks, which were excellent sellers, to help the historians stay in business. The Smithsonian held festivals which celebrated natural food. But at some moment in history, everything changed. The sheer numbers of people who needed to be fed overwhelmed every thought of the cultural aspects of food. Even the history of native American life was lost except for the few diehard visionaries. Too few understood or mourned the loss.

Too many people suffered from sensory overload. In a world overwhelmed by noise, by music, by digital games that created couch potatoes, cell phones, tweeting and twittering, pompadour preachers promising the moon and delivering thin air for the most part, people began not to hear. They seemed unaware of the loss. Even when thoughtful people informed them and they could see that nothing was left but frozen foods without substance, entropy set in and the people waited for someone else to solve the problem. There was no one to accomplish that. Even the international leaders in food safety did nothing.

FORTY

Maybe Live, Maybe Die

Out on the blacktop, Sam rode through the morning at top speed, angry, careless, and for one moment perhaps suicidal. But for some reason, he slowed the heavy machine and stopped beside the road. At last he revved the cycle into motion again and this time rode with purpose.

Sam topped a hill and slowed. Below he saw what once had been a picture-perfect farm. Long ago, someone who understood the land had planned it. A weedy acre and more that had been a garden, cultivated for the use of family and friends. The house itself was Prairie Style clearly patterned after Frank Lloyd Wright's architectural signature structures. Behind the spacious house, a large barn loomed over the space between house and yard. Someone had built a wooden deck that had seen too much rain and wind and needed staining. And everywhere were thick terracotta pots with the sad remains of flowers and herbs. No one had tended these in awhile, he thought. Beyond the house were open fields, where corn and wheat might have grown. There was an orchard whose tree limbs looked stick-like, bleak as winter, and there had been no fruit for some time. Would there be anything in the spring? He doubted it.

There was hardness in his eyes as he stared down at his family's home. Once down the hill, Sam stopped at the rusty mailbox and opened it. Pieces of envelopes and magazines spilled out, cobwebs covering them like a caul. He opened the pages of a seed catalog which contained the colorful legend, "ORDER TODAY. SUPPLIES LIMITED FOR NEWEST HYBRIDS!"

He reached down and discovered a tiny plastic spike lying on the ground. It identified "Husky Cherry Red hybrid tomato," with the last two words written in small six-point type which he could barely read. Sam frowned and let the catalog drop to the ground beside the cycle and then circled the house and at last stopped. "Anybody here?" But his voice was soft, and in it there was no hope of an answer.

He felt the wind whistling around the side of the house, and he opened the back door leading inside. He started to go in, but something stopped him. His hand remained on the screen door handle, but he could not pull it open. There was a sound that he could not identify, as if someone were breathing hard. He called out again. "Anybody here?" His answer was the wind. He held the door half open, and then he slammed it shut, as he had done so many times in his life. He backed away, and ran for the shelter of the barn.

The building was bare, with no evidence of life. No sound of pigeons or owls or cattle. Nothing. An old fishing pole leaned against the wall. He touched it.

Outside, Sam found a small grave, and on the brick, scribbled in white was the name "Baron" and along side it, a dog collar. Later, he leaned against the fence, a feeling of abject hopelessness filling him with its killing numbness. Tears slid from his eyes, which he wiped away. Funny how an entire planet was dying, but it was the death of a trusting animal that set the tears flowing.

He dug in his pocket and pulled out a photograph of himself much younger and with him was a small boy of maybe four or five. Both were holding fishing poles. Sam stared at it with such longing, but then the tears began, and he angrily stuffed it back in his pocket.

He had come so far, to discover this. There was nothing here but the ragged ruins of dreams and hard work. The hope that had kept him going through the long hard miles of his journey faded into the early twilight gray. Then he was riding away, somewhere, anywhere. Did it matter? But he knew it did, because he turned the cycle toward the town. He must find Colby.

Forty One

Choose Life

Near the two story white brick high school Colby stared up at the empty flag pole with its clanging flag hooks that flew no flags. Even if it did, what would they be, she wondered.

She stared at the front of the building, and started inside. Somehow she didn't expect to find much at the school, but her father specifically had sent her here, and she had nowhere else to explore.

Once inside, she caught her breath and practically drug her feet as she walked down the dark empty halls. Something about the silence made her feel as if she were about to jump out of her skin. When a loud b-r-r-inning cut through the dusty air, Colby spun around like a frightened cat. She had to find the source of the grating sound or get out of the building. The harsh clanging sound of the school bell, really a buzzer, bounced through the halls like a bullet. She gripped her club tightly and ran for the principal's office. Her ear drums were pulsating with pain. Gotta turn that off, she thought!

Nothing was in the welcome area, but she looked around and spotted an electronic apparatus above the receptionist's desk that controlled the bells. She studied it and quickly disengaged it. Who needed that, she thought. No one had responded to the call of the bells for months, and it appeared as if they never would again.

Out in the hallway again, it seemed an endless dark pathway leading to what Colby could not imagine. She glanced first at one classroom and then another. Each one had trappings of its subject. History and geography had huge maps, of the state, the nation, the entire known world, and then incredible shots from the series of Hubble telescopes.

Computers and monitors sat silent, with no young hands to engage them. She found the English room, and there were a dozen or so notebook readers, light as a feather designed to hold like a book while reading. They appeared to be no more than miniature laptops with silent gray windows, nothing showing on them. She made a mental note to investigate them further.

She passed a chemistry lab, and it showed signs of use. In fact, someone had apparently come in and wrecked it. Glass and tools were broken, spread upon the floor no doubt in frustration. But she was looking for the biology lab, which she finally found at the far end of the hallway.

The room was innocuous enough. At least no one had left any deceased biology instructors around, something for which she was profoundly grateful. She studied the room as she knew her father expected her to do.

Dusty books, discarded papers and pens cluttered the room. Who would have remembered to be neat on whatever final day was spent here? And did they know it was the final day?

She looked into a microscope, but the thick dust obscured anything that might've been there. That last day, she thought. What must it have been like? The last day of school, forever. The final meeting of friends, forever. Hard to think about. Those youngsters would've been experiencing the virus by then. And now, she thought, they were dead..

She walked past a row of glass containers on a shelf and glanced at their contents of seeds, labeled corn, beans, wheat, whatever was inside. She went on. There were dozens of empty pots with the rotted remains of plants. Was it the virus, or just that no one had watered them? She dismissed everything she saw. What secrets were here, she wondered. What did her father expect her to discover?

Later, still searching but losing hope, Colby stopped in front of the trophy case. Something caught her eye, the photograph of a handsome young quarterback posed as if about to throw the ball. She smiled wistfully, and then looked more closely. It was Sam! She stared at his young likeness, and felt anger flow through her. Why didn't he tell her who he was? This was his home and always had been. He must have family here, although they were all likely dead. Still, he should've told her.

What she failed to see was the white fiberglass cylinder hidden behind the photograph.

FORTY TWO

Dance A Little Longer

At the camp, Colby read beside the fire. As the light had disappeared, she dug in her knapsack for a huge hunters' flashlight, but then the cycle arrival brought her to her feet. She had no idea what to expect, didn't know where he had been, what he had done. She waited.

Sam came into the circle of light, sheepish as a boy caught in a prank.

"Sorry I was rude. Didn't mean to disappear on you. The idea of all this..."

"Overwhelmed you?" she finished for him.

"Yep."

Colby shrugged. "Me too. I'm reading these old cookbooks to keep from thinking about it. And I found this old seed list of native varieties."

Sam took it and began to thumb through it. "Beans. Blackeyed peas, chili peppers, plain old lima beans, gourds and greens, lentils, melons, okra, everything right here."

"I know," she said.

"Let's send for some," he said, and she smiled.

"If only we could," she said.

The awkward moment that followed lasted only a second and then Sam walked over and grabbed Colby's hand. She started to pull back, but something stopped her. Maybe it was time for her to trust someone, and maybe Sam was that someone. Besides, what was there to lose?

"Come on. I need cheering up!" he said.

She couldn't argue with that. She grabbed Iona and the flashlight and followed. It was a cold night, but not unpleasant, which was a favorite memory of the earth. While Polaris Space Station had "perfect" temperature, it was still artificial and could not improve upon the earth's climate and a night like this. Even the stars were brighter, and she had feared that their beauty had been lost to them in their engineered world.

Sam strode down Main Street like a football hero after the State championship, familiar, confident. He seemed to know where he was going, although there couldn't be that many places awaiting them. Soon, he stopped on Main Street in front of the general store.

"This looks like the only nightclub open," he said.

Colby shook her head, but the slightest smile escaped. "The last man on earth and you're a nut."

His response was to shrug and pull her through the door. She went willingly. Inside, the store has been cleaned out pretty much, although it did not look ransacked. The shelves were empty, except for odds and ends, laundry soap, paper goods, and cobwebs holding the carcasses of bugs. For a moment, their plan for fun faded as they stared at the emptiness.

Sam broke the silence. "They didn't leave much, did they?"

Colby ran her finger through the dust, leaving an "X." She set Iona on a counter. "Why should they leave anything?"

Sam picked up a plastic bottle of bleach. "You're right. Once the virus attacked the food chain, you ate the food and died or didn't eat the food and died."

"Once we know the virus is gone, Polaris teams can restart grain crops, herds of cattle, all of that," she said.

He looked as if he didn't believe her.

"Cattle herds?"

"They're test tube babies now, but they have the mamas up there," she said.

Sam nodded. "Amazing the scope of that place. I don't remember how it started, but it's grown like a suburb."

"Won't help us, I guess," she said halfheartedly. At least she knew her efforts in space would help someone, if there were any left besides them. Right now, that seemed a remote possibility.

Sam didn't respond but began opening cabinets, as if an answer might be hidden there. He seemed to be looking for something specific.

Colby leaned against a counter, patiently waiting for him to reveal what he was looking for. Her next question was idle, not entirely ingenuous.

"What does it do, the virus I mean?"

Sam dropped to his knees and looked under the counter. "Didn't you get relays, pictures?"

"Yeah. We saw pictures. Some athlete started the marathon in the Olympics, the one in St. Louis, and died of heart failure before he was half through. Of extreme old age. But we thought that was an anomaly."

Sam got to his feet and dusted off his pants "No. That wasn't an extreme. Remember a disease called progeria?"

Colby nodded her head. "Accelerated aging. A little kid would be the equivalent of 80 years old and he'd be maybe eight."

Almost without thinking, she picked up thread, a trowel, and pliers and tucked them into her knapsack. "A class I took, we watched an old show. Back in 2010, Barbara Walters did a special on three little girls with the disease, out of only about 45 in the world if you can imagine. They were looking for a cure among cancer preventing drugs, and they found it, about 2020."

"That's what this thing is like. The whole world has progeria. A genetic mistake. Your arteries clog and you get heart disease or something and you die at thirty or earlier."

Sam was studying the room, his interest in what Colby was asking wavering. "You are one fun girl. Let's change the subject. OK?"

Colby wasn't quite ready to let it go. "Maybe we're like some insect populations that live only a few hours or days."

Sam laughed. "No, of course not, but you could make a case for it. No one picked up on it. Like most things they do. They're the last to know."

"And the slowness was because the government was in charge. Talk about glacial movement," Sam added.

"We work for the government," Colby said.

"I rest my case," said Sam. "Look how it all turned out under our watch."

"Yeah," she said, the gloom resurfacing in her eyes.

Sam climbed into a large storage cabinet and dug around.

Colby seemed exasperated. "What are you looking for? We can't eat anything you find."

Sam's voice was muffled. "Hold your horses." With that he backed out of the cabinet, holding a slender green bottle.

Colby looks at it puzzled.

Sam looked pleased with himself. "Here it is."

What Sam had found still didn't register. "What?"

"Scotch whiskey. Mr. Rose kept it hidden. People around here didn't believe in drinking...in front of each other," he said.

"Is it safe?" she asked.

Sam checked the seal. "Was booze ever safe? But, yeah. It's old enough, and the seal's solid. Care for a drop?"

Once again, it occurred to Colby how crazy the situation was and how little control she had. "Don't mind if I do."

He found paper cups and poured out generous drinks. He handed Colby hers and held his cup in a toast. "Cheers."

"Cheers to you, too," she said uncertainly, and took a big gulp and almost coughed it back up. Her throat burned for a second.

"Ahhhhh." she whispered.

Sam grinned. "This is what they call sippin' whiskey. Smooth, isn't it?"

Colby made a terrible face. "Is it?"

But then in a minute, she held out her glass for another drink, and Sam obliged.

"I take it you're not a regular drinker," he said.

"Not allowed in space, so it's been a while."

"You better take it easy then."

"Sure. No problem." She shrugged. "Why?"

Sam shrugged. "No reason, I guess."

While Colby sipped her drink, Sam nosed around and found a CD. He popped it into Iona.

"Shall we dance?" He offered the girl his hand.

Iona obliged, and the tune emerged, *Our Love is Here to Stay*.

"Never heard that one before. I like it," he said. The music played and Sam danced Colby around the room. She was a little stiff at first, and then he held her closer. She surprised herself at how much she was enjoying it, and then Sam stopped dancing and leaned down to kiss her. She seemed to melt into his arms, and the kiss felt warm and lovely, but then her scientific nature lead her to stop the kiss abruptly.

"You're kissing me, and I hardly know you."

Sam started laughing and then spoke. "What difference does it make if you know me?"

Colby pushed away from him. "I don't know a thing about you besides your time in the Antarctic. What did you do when you were young?"

He shook his head in disbelief. The only girl in the world and she's a prude. "I told you. I'm a failed horticulturist. I studied crops and why they failed in different climates. I've been four years in Antarctica ...and the only female there was my boss, a career woman. About sixty."

Colby backed away from him, playing cat and mouse. "So you come after the first girl you see."

Sam was not as playful now.

"You're the only girl I've seen. You're the only girl I'm ever going to see."

Her voice was ironic. "I may not be your type."

"Who cares if you're my type."

Colby smiled and sidestepped away. "Or maybe I am."

"We have to think of the future of the world," he said lightly.

That remark sets Colby off. "The future of the world! I hadn't realized the depth of your concern for that. I thought the world didn't have a future."

"I bet you were a lot of fun in space. Let's drink to something."

"OK. Polaris and the future?"

"The past," he said.

Colby thought that over and decided it was a good plan.

"A fine idea. Drink to it, and then forget it."

"Ladies first."

"Lady first. OK. I propose a toast to my first bicycle."

Sam found that a little odd, but he drank to it anyway.

"To Colby's first bicycle."

"Now you."

"I toast catfish caught from the river."

"To catfish. May we see them again one day. We grow them on Polaris. We can import them once the virus dies."

Sam shook his head and spoke ironically. "Wow!"

Astrofish. That's what we'll call them. Fried astrofish at The Fishery. Your turn."

Colby thought a moment. "July 4th."

"Why?" he asked, like a teacher encouraging a student.

"Independence Day! Picnics and fireworks! Three kinds of ice cream and hot dogs," she added.

They both lifted their glasses to July 4th.

"Brothers and sisters," said Sam.

"I never had either, but I'll drink to them," she said.

"First love."

She stared at him. "I never had that either."

He was staring at her, and then he put his glass down and walked toward her. She started to back away, but she knew that somehow, she did not want to. She wanted to be near him. He smelled of soap and leather.

"There's no time like the present." He pulled her to him, and this time she welcomed his kiss.

The time slid away, and then they were dancing again, and she was thinking as most young people think that this was her love, and that if there was no future, and no other chance for it, then why not give herself to him. But then she opened her eyes and glanced over Sam's shoulder.

FORTY THREE

Nightmare

The scream came unbidden, like a nightmare, like something chasing you, like running and getting nowhere. Peering through the store windows were the faces of a half a dozen very old people, and they were angry and accusing, as if the young ones had forgotten some promise.

Angel was in the forefront and he stared at them with hate and fear.

Colby pulled away from Sam and pointed at the vision in the window. "Ahhhh!"

Sam saw them too, and ran for the door.

"Wait!" he shouted. He burst through the door and was out on the sidewalk. Colby ducked behind a counter and waited. In a moment, he was back.

"No one's out there."

The mood was broken. Colby was stone cold sober, and she was concerned, afraid that the men would steal their food or Iona. "Let's go back to the camp."

Sam nodded and put his arm around her. Outside he hesitated. "Are you afraid to walk back?"

"No. If anything happens, you'll hear Iona."

"I'll be along in a while. I just want to look around."

Sam gave her a gentle kiss and walked away into the dark.

Colby felt herself shivering, but then she headed in the other direction.

Sam made his way down a side street, looking for the group. He thought they might be in an old warehouse he knew of, and as soon as he discovered the building he was looking for, he opened the front door. He shined his flashlight inside and caught the reflection of a line of shiny John Deere tractors.

"Please come out. I won't hurt you."

He heard the shuffling of feet, and turned in that direction. He strained to see, but his light illuminated only pieces of boxes. Outside, the wind turned colder and set up an eerie howl. He turned and left the building, disappointed.

———

Colby walked into the light of the lantern in camp, and put Iona inside the tent.

Like a child, Iona protested. "Colby, I want to stay outside with you and Sam. Colby!"

Colby spoke as if to a child. "It's your bedtime."

Iona was petulant. "I don't have a bedtime. You just want to be alone with Sam."

With that, Colby zipped the tent shut.

"You're a smart little robotic."

She picked up a bag, took it to the fire, sat down and began combing her hair. The sound of footsteps brought a smile to her face. "Sam? That didn't take long."

She glanced toward the figure approaching, but then Colby's face changed. The figure standing at the edge of the firelight was not Sam.

FORTY FOUR

Potter

There was something familiar about the large handsome man well over six feet tall standing a few feet away staring at her. The enigmatic smile on his face puzzled her. What did he want?

Potter was perhaps 40, graying, and he was wearing military-issue fatigues. He was the man they had spotted on the hillside, the man carrying binoculars, a canteen and a sidearm in a holster. The weapon appeared to be a .22 pistol.

Colby scrambled to her feet and began backing away from the stranger with the gun. "Who are you?"

He dropped his head and then looked up at her. It was as if he had to struggle to remember his name. "I'm Potter. Where's the other guy?"

Her answer felt lame. "I don't know. Probably looking for you. He's...he's...Sam." It struck her that she did not know this person, had no idea if he was friend or enemy. Her voice went up an octave. "Sam!"

At that moment, Iona's siren went off, screaming through the night, like some terrible oracle of doom. On the outskirts of the downtown, Sam was walking slowly through the silent wood. He had almost given up looking for the Old Ones when the siren and Colby's voice brought him up short. He pivoted toward where he had walked, and then he was running back to her.

Iona continued to scream following her program perfectly. Potter and Colby remained on opposite sides of the campfire appraising one another. Colby eyed her cache of food, but it was the gun that troubled her.

"That's enough, Iona!" Abruptly, Iona shut off.

Sam came pounding into the camp site, his knife in hand. He saw Potter and he slid to a stop.

The appearance of the man surprised him. "Who are you? Where did you come from? Why..."

"Easy, Sam. Give him a minute. He's been watching us."

Sam didn't budge. "Why?"

Potter shuffled his feet like a child before a teacher. "I was scared."

Sam looked the man up and down. The fatigues drew his interest. "Scared, dressed like that?"

The man touched his shirt and then his pants with pride. "It's my uniform."

Sam seemed confused. "You're not regular army?"

"No. National Guard! I'm in the National Guard."

Sam seemed to relax at the comment. "Oh. She thought she saw some other people around here. So there are others still alive?"

Potter seemed unsure. "Not many. Everyone in town is dead. Some of the farm people lived longer. My daddy, he's ..."

Colby didn't let him finish. "Why was that, that the townspeople died first?" Colby asked.

Potter seemed eager to answer. "They had stored up food, canned stuff, that wasn't poisoned. But that ran out. No more trucks came to the super market. They're all gone, my teachers, the postman, the depot manager."

His eyes shift away from them, glancing toward the town. He is lying maybe.

Even so, Colby found herself feeling sorry for Potter. "Are you hungry?"

She had struck a chord. Potter's face lighted up.

"Yes, ma'm, but I have food."

He pulled out three potatoes that looked a little the worse for wear.

Colby eyed them warily, shaking her head 'no.' She tossed him a packet of food and Potter tore it open and gobbled it up. He dropped the empty package on the ground.

Sam gestured toward it. "Better pick it up. She likes things neat."

Potter obeyed, like a child caught making a mess.

"Now you're full, can we drop the games? What happened here?"

Potter was uneasy.

Colby saw it and pointed to a place near the fire. The large man sat down gratefully. "Let me get you coffee."

"I'll get it," Sam said and went inside the tent. In a minute, he was back with the hot coffee and packets of sugar. Potter took the sugar and poured three of the bags into the cup.

"My family sent me away. We had a farm and the grain blight came. They sent me to find a place where the food was OK."

"You didn't find that, I guess?"

"No, and when I came back, they told me to wait. They expected someone to come help us."

Sam's face seemed to darken, but Colby brightened.

"Meaning me?" she asked.

"Or him. I didn't know. Then I saw both of you," Potter said.

"Do you know anything about the two scientists up on Clay Street? They were killed," she said.

Potter shifted uneasily. His tone was evasive. "I don't know. They went away or maybe they died. But my papa told me somebody would come and help me. You must be the ones my papa told me about. How will you help me?"

Sam looked at Colby, but his expression was unhappy and his tone sarcastic.

"It's a giant treasure hunt. We're looking for the Fountain of Youth. Know where it is?"

Colby merely stared at him, not joining in his obvious cynicism. Potter looked hopeful, and then somehow realized it was a joke, at his expense. "Oh." His voice sounded disappointed.

"There's an extra bedroll. We all need some sleep."

Colby felt the tiny computer disk inside her shirt pocket again, but did not reveal it. She would keep it there until she could listen.

Iona began a slow version of *I'll Be Seeing You.* Why was she stuck on that, Colby wondered, just as she was slipping into a deep sleep.

Far above, on the dark hillside, Angel and the other survivors stood silhouetted against the sky, staring down at the camp. Angel was making his decisions.

Forty Five

Jonas

One Year Earlier

Potter was eager to tell Sam and Colby what he knew. Three stories seemed particularly relevant, once Potter revealed the devastation that had destroyed Central Texas. He didn't seem to have any idea about the rest of the country.

The first was about Jonas Burlington who Potter said was his cousin. Jonas grew the most productive garden in town.

Long before the blight struck, Jonas was an "old person" according to his grandchildren who lived a few blocks north near the high school, but he still thought of himself, a widower, as a virile and desirable man. He owned a ring that he sometimes showed his family. The ring, something he had purchased in Dubai when he had worked for an oil company, had belonged to his wife, but she had rarely worn it. Arthritis had prevented that and he had left it in his lock box at the bank since her death seven years earlier.

One evening, at the community center, Jonas had stood in the middle of a group of graying party goers, all gray that is, except for young Potter who still had dark hair. Jonas smiled at Cora Rogers whose shiny curly gray hair attracted him like a bee to a flower. He held up one finger in a gesture that suggested for her to wait a moment before she fetched another margarita for him. She narrowed her eyes in puzzlement.

Suddenly Jonas dropped to one knee.

The room fell silent, expectant. Even the elderly band leader halted his endless chatter and stared. One of the band members, a guitarist, the youngest person at the gathering at twenty-nine, nudged his neighbor, a bass fiddle man, and urged him to look at the old guy down on one knee.

"My dear Cora," Jonas began." I offer you this opal ring as a sign of my love. I want you to have it, and move into my condo and be my wife," he finished.

Cora's eyes widened in surprise. She appeared to be looking for a place to hide.

"Eh, I think I won't," she said.

Jonas was thunderstruck.

"Let's just remain special friends," she offered.

"You won't?" Jonas could not believe it. The band leader decided it was time to start the music again, and the imitative sound of Harry Connick Jr.'s big band filled the room. The partygoers started to dance, leaving a puzzled Jonas kneeling beside the bandstand.

"We'll still dance every Saturday night," Cora offered by way of appeasement.

Jonas begun to try to rise from his kneeling position. He could not believe her answer. How unreasonable, after he had made such an effort.

"Isn't the ring nice enough?" he asked.

Cora glanced at it again. It was a blur in his hand because she was not wearing her glasses. "It's not that."

"I exercise and I take vitamins," he said, his voice now plaintive. He was struggling now, attempting to rise. Two friends came over and each took an arm. "Come on, Jonas," one said.

Cora frowned. "Couldn't you leave things be?" she asked.

"No," he said.

"All right then, I'll say it. We're too old"

Potter stared at the pair. They both seemed very old to his young eyes.

Jonas looked at his lady love.

"It's never too late," he muttered to himself. He was dead wrong about that. In another week, he found himself falling unexpectedly. Within two months, he was bedfast. In six months his friends were attending his funeral. He appeared to be a very old man. No one knew why. Not yet.

Forty Six

Angela's Mirror

One Year Earlier – Hollywood

Angela Azmos was named the most beautiful woman in the world by her media team, who then convinced the papers and television celebrity shows that she truly was. They compared her to former beauties from all over the world, the late Elizabeth Taylor, Catherine DeNueve and Catherine Zeta-Jones, and the current Miss Rio de Janeiro. Seeing her on the Red Carpet at the Oscar presentations made it difficult to argue with the notion.

At age thirty, she was at her peak. Perfect skin. A slightly restructured nose which added an aristocratic look. A slender liposuction-enhanced figure that was quite voluptuous. She'd had no children although in those days, with gyms on every corner and exercise classes on every television channel, it would be difficult to detect one way or the other. Hands, fingers, feet, toes, eyes, eyebrows, even her ears had been pinned slightly. All was right about Angela. Any paparazzi would confirm that notion, particularly when they were running down city streets after her to get the perfect image for the television shows that loved to feature her.

The discovery happened about 7 a.m. on a summer's morning in Los Angeles. Angela was scheduled for a touch-up from her plastic surgeon. She had spotted a thin line on her forehead that stubbornly stayed in place despite efforts to fluff it out with a strong product her surgeon had prescribed.

By eleven a.m. she was ushered straight into Dr. Menas' office. No waiting for the international beauty. The doctor had studied her face and reassured her. After she had gone, he wondered aloud if she were actually thirty, or perhaps a bit older.

Later, Angela sat in the backseat of her limo studying her forehead as if it were a map, which it was beginning to look like. Ralph sat in the driver's seat waiting for orders.

"My beauty is fleeting," she whispered. "It's almost gone."

"Oh, no, Madame. You are ever beautiful."

But he found himself staring at her in the rearview mirror. Something had definitely changed. "There are excellent parts for older..." He stopped. What was he saying? He was out of place to comment at all.

"Take me home," she said. That afternoon, a staff member found her huddled in her bed, silently weeping.

The next day, Angela got into her most stylish clothes. The makeup artist would be waiting at the studio. She had used the same one for seven years, and she would not change now.

Within three hours - the makeup had taken a little longer than anyone expected - the screen test had begun.

The Director was well-known and well-loved for directing beautiful women as if he cared deeply about them. He assured Angela that the test was not his idea, but that the suits had insisted. It was something to do with insurance, he lied. What it was, a production head had run into Angela at a party over the weekend, and he had noticed a line on her forehead. That was the real reason.

The test took about an hour. The digital technology captured the event so that the Director and the Production Head, who had shown up unannounced, could see the results right away. What they were seeing was untouched, but would help them make a decision about the beauty's ability to play the lead in a billion dollar film.

Angela had insisted on her usual lighting man, and of course the makeup artist who had painted her face for years. The first medium shots were good, nothing unusual, but the last five minutes of the film showed something else, something unexpected.

The Director was frowning as he stared at the monitor. Angela peeped over his shoulder. "You have some mascara that's smeared, on your forehead. The makeup artist rushed over and began working on the visible line which she knew and Angela knew was not mascara. Angela could see exactly what she saw in her bedroom, except to her critical eye, it was deeper than it had been. She signaled for Mary E. to come over and look. Her assistant stared at the image, and got a strange look on her face. At that moment, Angela knew. This time there were no tears.

"Call me at home," she told the Director.

"I will," he promised. But he never did. He had his own problems, a severely sprained back and knee suffered in a handball game, the diagnosis advanced arthritis which was something new to him.

By Friday, Angela was officially written out of the film. She cursed everyone in sight, and wrecked her dressing room, breaking every mirror in the small trailer. By the weekend, she was forty.

FORTY SEVEN

America's Team

It was football season, a year earlier. The Texas UIL state football championship was on the line between the two best high school teams, both with outstanding records in the tough 6A competition. The stadium held a record number of attendees for a high school game, nearly 60,000 fans, many from other schools that had lost to the two teams.

On the field at Cowboy Stadium in Arlington, Texas, Leland Montgomery took two steps back and let the pass fly. His coaches were screaming at him, from the sidelines, from the grandstands, but for some reason he could not hear them fully. The sound, while at full decibel strength for almost everyone else at the game, was somehow muffled for Leland and that was before he inserted the inner helmet microphone that allowed him to communicate with his coach. .

What he did hear, as if from a distance, was someone saying run it, run it, but now the ball was up there flying across the yards, and his favorite target Zeb would catch it on the ten yard line, possibly take it in for the score which would give them a one point lead.

Suddenly the low roar which he could barely hear fell into a strange stillness. He thought he heard something like the air hissing out of a giant ball like they had at the Macy's Christmas Parade in New York. "Whoosh!"

The crowd as one looked frozen. Leland couldn't understand it. But Zeb was trotting back to him, a puzzled expression on his face. The umpire picked up the football which clearly Zeb had missed.

"Geez, Leland, the ten. Not the thirty. You threw it to the thirty," Zeb said. "You threw it twenty yards behind me."

Leland examined his large quarterback hands.

"Must've slipped" he muttered.

Zeb did not smile. "Must have," he agreed.

The following week the representatives from the various colleges would be arriving on campus, and both Leland and Zeb would be announcing to the press their choices of schools.

When Leland threw for the scouts, what had been smiles and hopefulness, turned to disappointment, almost disbelief. The boy who could throw the football a country mile, couldn't even get it to the fifty. Inevitably word would get out that the young man was injured. No scholarships were offered and none would be.

They noticed it first, the athletes, movie stars, television news readers, body builders and others who relied on their physical looks or skills. Many of them scurried to their doctors, but the physicians had no idea, not yet. Even when they did know, it didn't change anything. They became as ill as their patients. It was far too late to counter the virus.

204

Forty Eight

Seeds For Tomorrow

What no one seemed to know was that the seed saving project, begun in the late 1990s, had the potential to save the earth. But it seemed that the world was caught up in itself, and didn't pay attention. Fortunately, there were always a few visionaries and Jake Addington was one of those. He realized early on that the seed project had much greater value than many had thought. He wanted to know everything he could about it because he also realized that he could duplicate some of the essential efforts on the space station. With that in mind, he planned an extended trip to Norway.

The Norwegians were visionaries, over the top in their enthusiasm for collecting the seeds, according to some. They had something that few Americans possessed, the patience to see into the future at least as far as any men could. Because of the geographical limitations of the country, its limited arable land, its location among the fjords, they understood a simple truth: you don't foul your own nest. It was something the Americans and the Russians, and the Chinese with their endless space, never quite grasped. But even they could run out of good earth. It was possible to ruin the bounty that they thought was theirs forever.

There were 2,000 seed banks in the world, from New Mexico to Decorah, Iowa, from Oslo to Colorado. There were three in Brazil, and one of the three was concentrating on the essential Brazilian rain forests.

Jake had been invited to return to the secret Norwegian cave that housed the Svalbard Global Seed Vault early on, and there he was initiated into an international project that would have the potential to save the world from a series of natural disasters. Of course, then, they had no idea what the potential for disaster was. Most thought of nuclear winters after a war or a terrorist attack. They envisioned that a nation might be destroyed and they could restart its agronomy.

Jake had gotten there on a summer's day, with an expansive periwinkle sky that gave no hint of what the harsh winters could bring. He had come directly from Fort Collins, Colorado, and the U.S. National Seed Storage Laboratory. What the two seed banks had in common was that they were designed to withstand powerful earthquakes or even a nuclear disaster.

Collected there, 425 feet into the frozen Rocky Mountains, under an unassuming flat stone building, was every seed that had some economic relevance to life on earth. For years, the seeds had been frozen, but in later years a different kind of freezing called cryogenics had come into use. That procedure allowed the scientists to store seeds at low temperatures of approximately 340 degrees below zero. In that environment, the scientists claimed, the samples would last more than a thousand years. Jake was fascinated when his guide showed him how the procedure worked.

The samples were stored in thin plastic tubes, which were then sealed. Tall metal boxes held them within a round stainless steel vat. Whether the universal corn or wheat or pumpkins or squash, watermelon, cantaloupe, avocado, you name it, it was there. Every seed was carefully labeled including variations of any type plant such as tomato. Of those there were hundreds of varieties.

In all, by 2010, the laboratory had collected more than 350,000 subject types of seeds. The scientists were confident that they were protected from famine or nuclear holocaust, even plague or some environmental blight. Years later, the virus was a surprise and it was the exception that proved the rule. The seeds were safe, but almost impossible to reach because of a security system that worked.

A few years later, Jake discovered that the collection had grown exponentially. The scientists there had collected more than a million seeds and had gathered them in less than ten years. One of the scientists called the massive collection the "doomsday seed vault." Even so, half of the world's seeds had not yet been collected despite nonstop efforts by scientists all over the world.

Jake joked that it was too far to come, 620 miles from the North Pole, to be in any danger from enemies. His guide noted. "That's the beauty of it. No one will come after this place."

Jake had to agree reflecting on his own long journey to Oslo and then up to the Svalbard archipelago in the Arctic. Truth was, seed banks could be destroyed more easily than one would suspect.

The seed banks were quietly growing around the globe.

In the Philippines, a typhoon destroyed a fledgling seed bank, and they had not restarted the project. Lengthy and ill-advised wars in Afghanistan and Iraq devastated seed repositories in those countries. The devastation to the population and the infrastructure was so great that the last thing scientists were thinking about were the seeds that made the food supply. That was their mistake.

The bank in Norway had only recently, in the past five years, received Ethiopia's collection. That government feared it could not protect its unusual seeds and asked for secret escorts from the United Nations to help them transport the seeds to what they considered the safety of the Norwegian caves.

Soon after Jake was named a director of the Svalbard Seed Bank, he received orders sending him to command the Polaris Space Station. His assignment was to take along at least five hundred samples of the seeds and see what would happen in the different atmosphere. He added animal semen and a handful of mother cows, goats, llamas and other animals he considered necessary. He was excited about the project.

Forty Nine

Destination: Polaris

She was well aware of being a rookie, her uniforms crisp and decorations shiny as new pennies. She had never been this excited to have arrived at a destination. It was what she had signed on for at the Air Force Academy. It was the reason she had labored intently for a master's degree at Purdue. It was everything she wanted in a career. Her Mom and Dad were scientists as well, and they were pleased that she had achieved so much at a young age.

She knew a great deal about the terrible tragedy involving Polaris when the newly redesigned shuttle Sunrise had exploded in 2040, but she truly felt that safety precautions were much tighter now, and she meant to go. It was the career she had chosen after all.

Her father's parting words were, "I wish I were flying with you. I wish we could go on this journey with you, but you go for us. You are the future." Little did he know how frighteningly accurate his words were.

She felt acclimated in only a few days. The time at Houston's NASA facility had prepared her for the weightlessness, the need for daily exercise, eating right. It came easily to her, because she believed in her mission which was to join Colonel Addington's team in studying the effects of space on plant development. There were pressing reasons for the study.

What she had not anticipated was the diversity and compactness of the station. It was like a small perfectly designed city with no wasted space. There was advanced transportation that ran through tubes of light. Medical facilities with the latest technology. Entertainment consisting of football teams in monstrous stadiums, soccer and basketball. A golf course. Bicycle courses. Restaurants. Movie theaters. You name it, the creature comforts were available to the young residents, most of them inspired by colleges in the U.S..

Years before, when NASA talked about sending astronauts and scientists to Mars and not bringing them home, the world had been horrified. Soon after, NASA discovered a way to assure that they would come home. But this place was large enough and cosmopolitan enough to be one's final destination. Of course, Colby wanted to go on to Mars.

The first day, Colby reported to the commander. Light streamed through the louvered blinds designed to give the illusion of being on earth. Long shafts of sunlight made bars on the floor and desk and sparkled on the shiny modern chairs and athletic trophies that lined the shelves. They were holograms, she had learned, because space travelers traveled light. At his desk, his head bent over a tiny KindleLite, Colonel Jake Addington did not acknowledge her at first.

"Sir?" she began. "Lieutenant Ryder reporting for duty." She held her salute until he looked up and returned the salute.

"Yes, Lieutenant Ryder. Welcome," he said, and gestured for her to take a seat. She removed her cap and sat down.

"Everything in order? You settled in?" he asked.

She smiled and thought of her Spartan quarters, and remembered her father's book-filled office where his cluttered research papers filled every shelf and teetered precariously. She loved her father's cave, but she didn't want to dwell on that now. There were more important things pending. "I am, sir, and ready for my assignment."

The colonel pointed to a miniscule flash drive that lay on his desk. "Begin by reacquainting yourself with the food challenge. Every article we could find, news reports, scientific studies, events surrounding the destruction of seed banks, everything is here. And more is coming in daily. I want you to become our resident expert on what may be a threatened food supply. I will expect reports of everything you cover. The second part of the assignment is lab work. I want you to oversee and report on all developments, and these will go to the White House, the Senate, the Department of the Interior, and the Federal Emergency Management Agency. They will no doubt wind up at the United Nations. It appears that the concerns are more that surface. Something is happening, and we have to find answers for the government."

"Or else?"

His brow furrowed and he hesitated before he spoke again. His voice was soft and grave. "We won't have an earth to return to. It's that serious."

She blinked. "You're saying life and death hangs on what we do here."

"Yes. It is possible. Unfortunately. There are other efforts going on, but there have been some serious viruses, one that hit India has been unleashed all over the world."

"Millions have died there, but is it what I've heard?"

"Tell me about that."

Colby frowned and then began to speak. "All it took was a very simple mistake in cell division and it happened with one of the Super growers. The plants that resulted triggered signs of physical deterioration, like someone was aging. They didn't even make it to the ship docks to be transported to the United States, is what I understood. Of course we knew that there were risks. Unacceptable levels of toxins. There was even the possibility, ignored for the most part, that pollen from engineered plants might fall on weeds and give them extra qualities, turning them into a super weed that would then have to be controlled."

"You heard right." The Commander's face showed his deep concern. "The world's food supply was at risk, we knew that. At first there were whispers, and then over a period of a decade, the voices grew louder. They're a shout now."

"Can it be stopped?"

"The study came out of MIT. They had a cure, or almost a cure, for Alzheimer's. They'd figured out that the 'mistake' caused DNA to replicate itself and clog normal cell development, which then led to accelerated aging. You've heard it called progeria, but that was different. This one hits anyone who consumes the hybrid products, from fruits, vegetables, alfalfa seed, beef, poultry, lamb, even fish, once the products get into the ocean."

"We have to have a food supply for survivors on earth?"

"Yes, the ones who were immune or who didn't partake of the usual foods, and we will. What we are growing here will be enough for a few."

It never occurred to Colby that she would need the new food supply. She'd always considered herself immune from the usual civilian needs, since she was part of the military.

FIFTY

Out Among The Stars

In her quarters, the air was cool and clean, thanks to the excellent air conditioning system that kept an optimal temperature at all times. Without it, they wouldn't be able to stay because of the extreme cold outside. With it, they could carry out scientific experiments, live a reasonably normal life, enjoy the scenery which was spectacular, and on occasion forget that they were thousands of miles from their home, the earth.

The crew worked thirteen, fourteen hours, stopping for meals, falling into their beds at night. There was little time for conversation, recreation, relationships. Their goal was too important, and leading them all was Colonel Addington. One day, Jake called a halt to the long hours, insisted that everyone take some break time. He wound up in the viewing room with Colby.

In the Space Viewing room, there was a computer programmed binocular, specially designed for the purpose of seeing the earth, even individuals at their tasks. They were similar to the venerable satellite cameras with their magnification capabilities.

Colby was still in awe of what they enabled her to see, when Jake arrived and insisted on giving her the tour of earth he had worked out over a period of years. First he showed her the North American continent and the various places he had been stationed. She showed him her stations too, but there weren't many of those. Then he expanded the view and included Norway, China and India, where his stay had been brief. India had been a fact finding mission, and the virus had already started there. Soon after, he returned for the second time to Polaris. This time he knew he would stay until he had answers.

Most of all he spoke of his time in Norway. He had learned everything about the seed cave, and he had made good friends with the staff there. He was amazed that they had an excellent grasp of the American language. When he commented on it, one of the staff laughed and said, "No one learns Norwegian, so we must learn English, German, whatever is useful."

It was a generous attitude, and it was true.

Jake was complimentary of some of the work elsewhere, and expressed his hope that it would bear results soon. He wasn't in a competition with others. The reality was that his antagonist was death. There was no second place finisher.

Once Colby realized what her task was, the work resumed. Her team was already at work, and she threw herself into the research, hardly stopping for sleep or meals.

She did take time to speak with her parents, and it was a talk with her Dad that left her shaken. He seemed hesitant to tell her, even though he knew that she was not his little girl, but rather a leader in the effort to find solutions to the growing threat.

"There's been some violence," Professor Ryder had said during their brief conversation.

"What kind of violence?" Colby pressed.

"In Peru," he said. "A Maoist group. Guerrillas, they're remnants of the old Shining Path. I think, anyway, they attacked the International Food Discovery Center at an agricultural research station in Huancayo, Peru. Some of the workers were murdered. It's no longer safe. The guerillas just thought they were gathering food for themselves. They had no idea of the research. Anyway, the Center is moving to the city, where they can guard the research."

"Is there more?" She was worried now, both about the seed banks, but about her parents as well.

"Four other potential seed banks are under fire. They're all tasked to preserve genetic diversity," he said.

"What happens if the research is lost?" she wanted to know.

"We're already seeing it. India was first. Now we hear Venezuelan, Columbian and Bolivian seed banks have been attacked. As far as I know the ones in the U.S. are protected. If you hear more about this, let me know."

"I will, Dad."

It was only a few days before Colby heard more news, which was frightening. The stories were so stunning that the entire crew was forbidden from calling Earth. Ironically the crew was allowed to hear speeches from Congress about how safe the food supply was. The congressmen would pontificate about how everyone should calm down, that there was nothing to fear, but then they would go to their special dining room and eat food that had been certified organic. There had already been warnings. The decline in genetic diversity would destroy the earth. There was a notable loss of life forms, rapid extinction of the germ plasma of food crops, even the basic food crops were imperiled, corn, wheat, rice, barley, millet and sorghum. Common sense was nowhere to be found as farmers narrowed the genetic strains. And once these were at risk, what happened was inevitable.

At some point in the evolution of the great United States, the governors, congressmen and women, Senators and the White House had begun to think of themselves as separated from the hoi polloi, in other words, everyone else on the planet.

That meant they gave themselves special favors in every aspect of their lives, food, transportation, living quarters, you name it. They particularly wanted to escape the blight that was overtaking the earth. So much for equality.

Still, Colby and Jake met to study the earth, its curvatures and lines of demarcation, its boundaries and forests and plains and particularly the great oceans. And finally the Hour of Shadows, as Colby called it. There was something sad about it, and yet there was majesty. To see all of it was magical.

It was during one of these moments that Jake knew he loved Colby, and that it was impossible. The military had long frowned on fraternization between officers and staff, particularly in foreign assignments, and one could not get much further than Polaris. What they meant was that sexual liaisons were against the rule. Jake understood why. He had seen commands devastated by such love affairs, by rapes, by untoward advances and he did not intend to let down his own standards even though he wanted to. Colby sensed the tension in him, and did her best to see him in the presence of others or not at all. She was too young to understand all that he knew, and he was too old to share it. The surprise for both of them was that time ran out on the possibilities. She was called home.

Now, huddled in the November cold, she didn't think she would ever see Jake again. And she had work to do. How incredibly inviting the space station seemed now, but that was another time.

FIFTY ONE

The Street Where You Lived

She was earthbound, and all that hopefulness involving Jake was behind her. She didn't think she would ever see him again, but she would not dwell on that. She had work to do.

Sam seemed to be searching for something as he strode down the street. There was a strained look on his face, anxious and alert.

Near the gazebo in the park, Iona sat on the ledge of what appears to be an old-fashioned 'wishing well.' As Sam approached, he was surprised that Colby was nowhere in sight.

But Sam's approach got Iona's attention. "Hello, Sam," she said fondly.

Sam hardly noticed. "Where's Colby?"

At that instant, Colby popped up from behind the well.

"What are you doing back there?" Sam asked.

Colby slipped a tiny computer disk like the previous ones into her pocket and dusted off her jeans. "Just checking out the well. If this water's not potable, we're in trouble."

Colby grasped the bucket, but didn't quite have the hang of it. Sam took it from her.

"Let the country boy do it."

Colby released it gladly. "Thanks."

Sam expertly dropped the bucket into the well and it landed with a splash that echoed up from the cavernous emptiness. He pulled hand over hand on the rope until the bucket reached the top and he balanced the container of cool blue water on the ledge.

"Got a test kit?" Sam asked.

Colby nodded her head and reached into a pocket for a tiny tablet. She dipped water into a container and dropped it in. Both Sam and Colby stared at it, the thoughts behind their stern expressions no doubt the same. Without clean water, there was no hope. When the water turned a dark blue, a look of relief crossed Colby's face first.

"We've got good water."

Sam breathed a sigh of relief, but then glanced down the street and spotted Potter walking toward them.

"And one uninvited guest."

Later, Colby finished searching the town and was walking back toward camp. A strange smell wafted through the air, the smell of French fries cooking. She strolled back toward the camp, reminding herself that she could eat nothing on the planet, at least not yet. There in a skillet were the most perfectly beautiful potato slices simmering in hot grease. A grinning Potter was expertly overseeing the cooking. Colby walked into the circle of light and stared at the delectable poison. "French fries! I haven't seen those in years!"

Potter grinned. "Found the potatoes awhile back in a root cellar. I been wanting to cook them." He dished them up, and offered some to Colby. "You're just in time."

They looked irresistible. Boiling corn oil, the carefully cut potatoes frying in it. Potter lifted out the strainer they were in and dumped them on brown paper. It had been years since she'd eaten anything like them, but she hadn't forgotten how delicious they could be. She reached out a hand, but then pulled back. The familiar words poured out. "They're probably not safe."

Potter didn't hesitate, but bolted them down. "I dunno. Maybe it won't matter much longer."

Colby pulled out some of her space food, and began eating it reluctantly. Compared to the potatoes, it was pretty boring stuff.

They sat in silence a moment, and then Colby's curiosity got the better of her. "Potter?"

"Yeah?"

"Did you see many of the people when they were dying? I mean, you must have. Your folks?"

"I saw some of 'em. Before I went away," he said.

She was surprised at how guileless he was, hardly seeming to mind talking about what must be a difficult subject. "Will you tell me how it was?"

Potter thought about it. Unnoticed, Sam stepped into the edge of the circle.

"Everyone 'round here farmed or ranched. At first, they talked about how this green revolution was a gift from God," he said, struggling not from emotion, but as if he were trying to remember.

Colby listened and then spoke. "We heard that in space."

Potter went on then, as if by rote. "On TV, they said it was a miracle. The whole world was gonna get fed. Called it the new grains, and the whole earth was blooming twice a year. The cattle were so fat!"

"Go on."

Potter seemed uncomfortable remembering.

"At school we heard a teacher telling how somebody was growing old before her eyes, and some others noticed it too, people looking old and then they were old, dying real fast."

Colby realized what he had seen and was horrified. "You saw that?"

Potter nodded his head 'yes.'

"The scientists on TV thought it was radiation. The ozone layer was disappearing. People was always scared of that. Then there were no more babies. Well, a few, but they were funny.

"Deformed?"

Potter looked confused, but then continued. "I guess. Then, everybody went kind of crazy. The farmers were sure it was a plot. They went after everybody they didn't know. There was killing."

Colby nodded. "Aging brains. Anger from what was happening to destroy the brain."

Sam stood listening, and then stepped in. "Let's change the subject. Cards?"

Colby sat back, thinking about the conversation while Sam got out his deck of cards.

"What'll we play for?"

Potter looked at Colby's backpack.

"Food," he said.

"How about ..." Colby looked around and spotted what she wanted. "...white rocks?"

Iona was singing loudly, "I'm getting married in the morning. Ding dong, the bells are gonna chime. I'm gettin' married in the morning. Get me to the church..."

The card game went on a while. Potter couldn't seem to keep up with a simple game of blackjack, and then poker, but he tried. Sam wound up with a stack of white rocks in front of him, and grinned at his adversaries.

By midnight, they had all gone to sleep.

Fifty Two

Poor Potter

In the days to come, Potter begged Colby to tell him about Polaris, and she surprised herself by taking a great deal of pride in the awesome invention man had contributed to the heavens. She told him something of the history of the research facility, and how international astronauts had gradually built the structure so that it could accommodate scientists from many countries, and their experiments in biology, physics, astronomy, meteorology and more.

Sam was listening as carefully as Potter, but pretended disinterest. She explained how Polaris had unusual conditions that broadened the fields of research to space medicine, physical aspects of going into deeper space, astronomy, which included monitoring Hubble XX, which continued to send back astounding images, and finally use by the private sector which had started only about five years earlier, but already had resulted in a number of inventions that seemed to benefit mankind.

Colby stared at the sky and then she saw it. "Look up there," she said to Potter. "Polaris. At one o'clock."

Potter's eyes followed where she was pointing. "I see it!" he said.

'That's it, she said. "Looks so close we could fly there."

"If we could fly," he said, sadly.

She felt sorry for him, but she was unsure why. Likely, it was never to be, but she refused to destroy his dream. "Maybe one day," she offered.

He stared at the bright object flying above, going thousands of miles an hour and yet it appeared to be moving slowly, almost floating in space.

The sad thing was that right now, there was no traffic back and forth from the earth. Potter told her he had always wanted to go there, so that he could see an "earthrise," something he had heard on the Discovery Channel until it went off the air.

After a while Potter's eyes began to close and Colby told him they would continue the conversation later. He willingly went to bed on a cot he had dragged into camp, and she soon disappeared into the tent.

At least their exhaustion enabled them to sleep late. In the morning, Sam pretended to sleep, but he was watching. He saw Colby's hand hit a button and pop out a miniature computer disk. It had been a lot of years since he had seen one of those. Now everyone used thumbnail sized flash drives.

She caught the drive with one swift movement, and then hit another switch, and a new song began.

"Come away with me, Lucille, in my merry Oldsmobile. Down the road of life we'll fly, automobubbling you and I."

Sam pretended to awaken. "What the heck was that?"

Iona twittered, pleased to be noticed. "My morning program. From the early 1900's, before your time, Sam."

"Before anybody's time."

"Sorry. She gets carried away," Colby said.

Sam was already thinking of something else.

"Where's our nocturnal visitor?"

"I sent him for water."

"You trust him?"

"I know as much about him as I do you."

Sam muttered something, but let the comment pass and poured coffee for himself. "Doesn't that gun he's packing worry you?"

"No. He seems like a nice farmer."

Sam took a sip of the hot coffee and blew on it. "If he's a farmer, I'm a ballet dancer. Last night, I shook his hand. Soft as a powder puff."

She frowned. "Nobody's farmed in a while."

"True, but even so, he'd still have some calluses on his hands." He hesitated. "When does this treasure hunt start?"

Colby bristled. It was much more than a treasure hunt, and she wished that he would admit it. "Now!" She picked up Iona and strode away.

Sam watched her a moment, and then, frustrated, he followed. He caught up with Colby in the middle of the street. She pointed toward the picturesque little church. "There."

"Oh, great. Sunday School." Sam looked bored.

"Nobody's forcing you to come."

Sam looked offended, and then Potter ran to Colby's side, his hair newly slicked down and his face clean, as if he were ready to go to Sunday School. Sam looked suspicious.

"I better attend. He's here," Sam said.

Colby elaborately ignored him and headed off down the road toward the church.

In front of the church, Colby stopped long enough to stare pensively at the building, and then she steeled herself and marched up the steps. Potter ran ahead and opened the door for her. Sam watched from the bottom of the steps, and then followed the two inside.

Within, the church was silent. A stream of light cut through the broken ceiling, illuminating dust motes hanging in the air. The musty smell of a long unused building made them hold their noses until they acclimated to it. She headed up the aisle, wary as a snake in the sun. She picked up a hymn book and flipped through it, but put it back down as Sam and Potter came noisily up the aisle. She stopped at the altar and stood staring up at the cross.

Potter picked up an offering plate and handed it to Sam, a slight smirk on his face as if he were waiting for something.

Sam took quick note of what he was doing, and joined in the game by flipping a coin into the plate. Potter looked at the coin in delight, childlike, and grabbed it, stuffing it into his own pocket. The two men soon grew tired of the game, and watched silently as Colby slowly made her way up a side aisle, ignoring them. She found a door, opened it and looked up.

A spiral staircase twisted its way up into the steeple, the cobwebs curled around it like crochet thread.

The two men watched her go inside, and Potter waited until she disappeared to pull out his pistol and twirl it. In a minute, he went toward the altar and pretended to fan the gun in that direction. Sam gave him a look, and then gestured for Potter to follow him outside. Potter holstered his gun and went outside.

Colby slowly wound her way up the dusty steps.

Outside Potter sat on the steps playing with his pistol while Sam paced.

"What's she looking for? They sure hid it good. What is it, Sam?"

"I don't know. Cache of food. Maybe."

"If you didn't come with her, why are you here?"

"Can you suggest some better place?"

Potter missed the intrinsic bitterness. "Than here? Sure. There's a lot of places. Dallas and Fort Worth. They have rodeos!"

Sam gave him a look. "Not anymore."

Potter looked disappointed and then cast around for something else to think about. "Would you show me how to ride your cycle?"

Sam was surprised.

"You never learned to ride a cycle?"

Potter dropped his head, as if ashamed, and then brightened. "I can drive a tractor."

"That's a handy thing to know. You learn that early around here." Sam was about to say more when the pealing of the church bells caused them both to jump to their feet and stare up at the tower.

"She's found something!"

"Whatever 'something' is."

Sam dusted his hands off. "Treasure hunt over!"

The bells slowed, echoed and stopped. In a moment, an ecstatic Colby ran out the front door of the church and down the high steps. She waved a tiny computer disk, like the other ones. She allowed them to see it, and then tucked it into her pocket like a kid hiding the last lollipop.

"I got it! Just where Dad said it was."

Sam left off his gloomy look for a brief moment, and Potter beamed. "Maybe there's hope for this enterprise after all," Sam mused.

Potter was interested, like a puppy after a treat. "What does it say?"

"Look, let me hear it alone. Then, we'll talk. This is the last lead."

Potter nodded agreeably, but Sam turned from them, clearly angry. Colby started away, and Sam grabbed her arm. She turned to protest and Potter wrapped his arms bear like around Sam. Sam shrugged him off and shoved the bigger man down hard.

Colby ignored the scuffle and walked quickly away.

"Stay out of this," Sam said to Potter, and then went chasing after Colby. He caught her and jerked her roughly around toward him. "When do you let me in on your secret?"

She pulled away. "When I'm ready."

"You better start trusting somebody soon!"

Colby stared at him, almost relenting, but something held her back. "My father...warned me. Trust no one."

"And look where his advice got him."

Colby reacted as if she had been struck.

—•—

Later, in the tent, with great stealth, Colby inserted the last computer disk into Iona and waited. Her father's voice brought her to full focus.

"This is our last message. Go to the County Seat, to the newspaper office. Look up April 24, 2020. This will lead you to the life source. I must go now. They are coming."

She did not dwell on what she knew had happened next, but instead grabbed Iona and headed toward her four-wheeler.

Iona's lights were whirring and blinking. "Is Sam going, Colby?"

Colby hesitated, a look of concern on her face. "No."

Outside, Colby counted out food packets. She was surprised to hear Sam and Potter stroll back into camp. She could see that Sam was barely controlling his anger. Strangely, he talked to the machine.

"Tell me where she's going."

Colby glared at him.

"I am not authorized. Sorry," said Iona.

Sam turned to Colby. "Where are you going?"

She ignored Sam, and finished her task.

Potter stood back watching, his eyes darting from one to the other. He searched around for an electronic chess board and placed it in front of Iona. "Let's play."

Potter searches for the proper button, pushes one.

The two concentrated on the game, while Colby and Sam continued their tense discussion.

"At least tell me which way you're going."

At last, Colby irritably pointed toward the east. "That way."

"Same direction you came in?"

Colby gave him a look. "Yes."

Sam's voice was accusing. "You're going back to Polaris. They said you could come back! That's it. I knew it!"

"I wish it were true, but no such luck. I'm stuck here with you!" she shouted.

"I'm going with you. What if something happens to you? What if something happens to the ATV? What if you can't fix it? What if...?"

"I appreciate your need to be needed, but I came a lot of miles to get here. I can handle ten miles to the county seat."

Sam was beyond exasperated. He was furious.

Colby was a little surprised at his anger.

"Your way or the highway! How typical of....,"

"I know! Of a woman! You had to say it, didn't you!" She turned away, collecting her thoughts, holding her anger in check. She had learned that in the military. She wouldn't have made it had she not. "If I find this life source, whatever it is, there'll be work to do. For all of us."

Sam's eyes remained hard. "Get him to do it."

Colby knew she was not connecting. She tried another approach. "If you think I'm not coming back, that's crazy. I don't intend to spend the rest of my life wandering around Texas alone."

Potter had seemed not to be listening. He made a chess move. "Leave her alone. She can do it by herself," he said.

Sam seemed to deflate then. He took a step backward. "Yeah. She can, can't she," he relented.

Colby slipped her knapsack onto her shoulders, and picked up Iona. "You can finish the game later."

With that, she climbed onto the 4-wheeler, started it, and rolled slowly away, gathering speed and kicking up clouds of dust. Sam watched after her, torn between fury and worry. Potter pointed to the board.

"Want to play?"

Sam stared at him, the fierce anger dying within him. Then he sat down to play chess with Potter. It quickly deteriorated into a game of checkers.

Fifty Three

Colby Races The Clock

Out on the cold blacktop, Colby turned the engine up a notch. There was something exhilarating about sweeping along the empty highway, the wind, the illusion of freedom, all of it felt good.

From a distance, high atop one of the gentle hills that protected the town, binoculars focused on her. Two shaking hands held them up to fading eyes.

Colby slowed for the railroad crossing and then halted the vehicle. Something had moved in the old box car that once was used for storing hay. She rode as near the car as she could and then got off, walking slowly toward the structure.

A slight movement caused her to glance toward the opening in the car. She strained to see, but no one was there.

She took a few more steps and called out, but silence enveloped her. She scanned the structure, the bales of hay and shrugged. With that, she returned to her journey.

Behind her, unseen, Angel and Dave watched her from inside the freight car. Angel lifted his rifle briefly, but thought better of it and lowered it again.

Fifty Four

Food Fight

One of the harsher policies some nations put into effect was to use food as a weapon. It was not uncommon, but it was cruel.

Only after Colby had long since departed for home, Jake remembered more of his research. There had been a story about a major agribusiness corporation, Monsanto, that had gone to court to ask that it be allowed to market something it called a "terminator" crop. Monsanto contended that it wanted to protect its intellectual property.

In this particular case, Monsanto had invested billions of dollars in developing bioengineered crops that were insect resistant, and that by spraying the plants with a chemical, could turn the engineered genes on and off, destroying the plants that had been genetically modified.

What the company wanted to do was splice plant toxin genes into crops that then would produce sterile seeds. The idea would be that new seeds would be required every year, and that would mean tremendous profits for the company annually.

Finally, the company bowed to pressure. Adversaries of the plan made it known that 1.5 billion people, most in poor countries, relied on seeds saved each year from the previous year's crops. People would starve when it was not necessary. Monsanto did back down, but even so, dozens of other companies were investigating the same technology.

Of course they didn't mind experimenting on the poor.

Amazing, thought Jake, that people were so power mad that they would use food as a weapon, but it was done all the time. He had seen it in Somalia, Darfur, Sierra Leone and South Africa. He had seen it in Afghanistan and Pakistan, but rarely in the European Union.

In order to get a bag of rice, you had to join whatever armed force had control of the U.N. trucks sent for free to feed whoever needed it. But the rice was never free. And often it cost everything a young man had, his life, his family, his hope.

FIFTY FIVE

Settling In

Sam carried a large box into Colby's folks' house. Potter followed with a bigger load, but he struggled and set it down on the porch, breathing hard.

Sam gave him a look. "That goes in the kitchen. Can't you finish what you start?"

Sheepishly, Potter picked up his load and obediently took it inside. Once there, he grinned. "Food goes in the kitchen," he said.

Sam groaned. "Thank you, Heloise the Fourth," he said, sarcasm in his tone as he referred to a home economics advice columnist from the early part of the century.

Later that afternoon, Potter followed Sam along the street, irritating Sam, but he said nothing. It seemed strange, after so long alone, to have this large man with the brain of a child following him around the town.

Sam hesitated in front of the old grocery store. It was the same as it had been, but everything was faded. There were signs of events at the high school, a pancake supper, a football game, a car wash the cheerleaders were holding to raise money for cheerleader camp, a poster advertising oil for tractors, a "for sale" sign for a lawnmower. The windows were dusty and hard to see through. Sam started inside. Potter stayed right behind him like some pesky little brother.

Inside, the shelves were mostly empty, but there were items such as laundry soap, Skoal tobacco and shoe polish, which Sam picked up and examined. He tossed it back on the shelf, not surprised that no one would be polishing shoes during the last year or two. He found a brush for horses, gear to outfit a horse, saddle, bridle, martingale, and a lead rope. The meat counter was empty, and surprisingly it was clean. He didn't know the proprietor, but could imagine that he took great pride in his place of business. He might've even hoped that the horror on the earth would be reversed and he would be back in business. Right, Sam thought. He glanced back at Potter who had found an ax which he was examining with care. Sam told him to put it down, which he did.

In the back office, Sam found a roll-top desk. Inside was a dusty scrapbook. He opened it and studied a series of news clippings and yellowed photographs. In the most prominent picture, a semi-circle of men dressed in ties but without their jackets, stood, and one man held a capsule up high. Another of the men held a shovel with a ribbon tied to its handle.

"They buried some kind of time capsule here in Rice?"

Potter looked carefully at the picture. "I know where."

Sam wasn't listening. He went on. "Looks like the whole town's out there burying a time capsule."

Potter was insistent. "I know where they buried it. It has secrets in it."

"Wonder what...?" At that moment, Potter's words began to register. "Secrets? What secrets?"

"Life secrets," Daddy said.

"Where?!"

"In back of the City Hall is where they buried it."

Sam didn't wait for an explanation. Potter stammered a little but then followed Sam.

"Let's go dig it up."

"Wait! I want to help."

Sam was grinning now. "She's on a wild goose chase. We'll find it before she gets back! Ha!"

In the square of the county seat, Corsicana, Colby stopped before the Corsicana Sun newspaper office and hurried inside. The place was dark and filled with dust and cobwebs, but there were no bodies. She was glad of that. She prayed as she switched on a computer. She had no idea if there was power, but in a moment she was rewarded with a neat orange cursor and a desktop of choices. She clicked on a file cabinet.

In Rice, Sam stared at an empty hole in the ground, his face reflecting both anger and surprise.

Potter stared into the hole with Sam, uncertain whether to speak. "Somebody dug it up."

Sam was instantly suspicious. "Somebody?"

Potter grinned boyishly. "Me."

Sam closed his eyes. "Where is it?"

"I'll tell you if you'll play football with me."

"Chess. Football. You're nuts."

"Pleeeease," Potter whined.

Sam looked hard at Potter and then gave in. It could be a way to get answers. There didn't seem to be any other choices.

"Come on. I promised you a game. Then you tell me about it."

Eagerly Potter followed Sam, his big frame lumbering along trying to keep up.

Fifty Six

Disappointment

In Corsicana, an agitated and furious Colby raced out of the newspaper office, jumped on her ATV and sped toward Rice. Was nothing ever going to go right about this search, she wondered, and answered her own question with a resounding 'no' that echoed in the streets and blended with the whistling wind.

FIFTY SEVEN

The Game's On

A football spiraled through the air, and into and out of the hands of Potter, who seemed upset at dropping it. He picked it up, and held it tightly, liked a stuffed animal.

Sam went out for a pass. "Now. Throw it now!" Sam urged.

Potter cocked his arm and threw, but the movement was awkward and the football dropped far short and wide of its target.

As he watched it go astray, Sam was a little puzzled at the weak throw.

Sam trotted over and retrieved it, and then came down to Potter.

"Timing's everything. Football not your game?"

"Never got to play," Potter said.

"A big guy like you? Surprised the pros didn't get you. You're a born linebacker."

Potter grinned happily. "You think so?"

"Sure. I was a quarterback." He hunched over as if waiting to receive the ball. "A long time ago."

Potter seemed enthusiastic. "I know. You won all those games!"

Sam looked suspiciously at Potter.

"How'd you know that?" Sam asked.

Potter reddened. "I saw your picture in the trophy case, in the school."

"No kidding. Still there, huh?"

Potter seemed to be thinking of something else then and stared off across the field.

"Do you think she'll come back?"

"Oh. Sure she will."

For a moment, Sam began to think of all that had happened, and he didn't notice that he was walking off with the football until Potter chased after him and caught him.

"Sam?"

"Yeah?"

"Leave the ball. I want to kick some field goals."

"Sure," he said, and then casually added, "The capsule you dug up. Where is it?"

Potter carefully placed the football on a tee. "In the trophy case."

Sam headed toward the old school.

———

Near the hay station, two of the Old Ones, Angel and David, stationed themselves behind a bale of hay. One of them lifted a deer rifle and rested it on the bale. Then they settled down to wait.

FIFTY EIGHT

Decision

The brain trust in Polaris had been challenged by the President of the United States to find an answer to the virus run amok on the earth. At the very least he expected them to figure out some approach scientists could take in fighting the destroyer. It was a challenge the Polaris leader did not take lightly and one he would not let rest.

Jake Addington had always heard he was too big to be an astronaut, at least according to the accepted way of thinking. What the opinion makers didn't count on was Jake's sheer determination. He was soon a Lieutenant Colonel in the U.S. Air Force Astronaut Corps, and that came early in his career. He pretty much wrote his own ticket.

He had been at the controls now for nearly seven years, in charge of Polaris. He was lucky that the crew was congenial, had been from the beginning. That had been one of the first traditions of the effort. It was international, and everyone understood that when you lived in close quarters, you couldn't snarl at each other, not so anyone would notice anyway. Today, he was looking at the earth from the observation deck, an area made comfortable early on, and it had always been a joy to watch the changing shadows that made the globe beautiful. But now, what was once called the blue planet, seemed somehow altered.

Something about the turning ball reminded him of his childhood and the oversized globe his parents provided him. That and the starry night created by the laser show in his bedroom where he studied the names of the planets, the stars and galaxies. Wasn't his destiny set before he was twelve, and that choice to live in space had given him extra days of life even though he had not planned it.

Right now, he was preparing to speak with Colby, his star agronomist, the young person he had considered "most likely to succeed." Funny to think of that now, success. If staying alive counted, then he had not been wrong.

But he knew he needed to help her, and today he would find out what he could do. He did have an ace in the hole, something no one else had thought of. He could land water and food on earth to sustain her life. If there were others she would have time to find them, and then there would be hope. But first, he had to know that she was alive. Fortunately, Iona had been programmed to signal the beginning of their talk, but still Jake's voice came as a surprise.

"Colby. Colby, come in. This is Jake."

He waited in the silence of deep space, frustration dogging his patience.

While he waited, he could not forget the evolving of the virus, how it had come into the food chain with a stealth unseen since the great flu epidemics of the early 20th century. A few people had seen the dark side of the new way of growing food, and they had called for regulation of the new crops. They appeared before state governments and regional entities and finally the National Academy of Science where they held noisy protests, shouting their concerns against Academy doors locked like closed mines.

But no one would make a decision that meant anything. Too much lobbying power, too many hidden dollars into the pockets of Congress, sunk any chance for meaningful legislation concerning crops. Finally Jake did try to make a difference, and it cost him. When he dared to go outside the official line, the familiar one that refused to question if bioengineered food might not be safe, he was no longer asked for his opinion. Instead, he was ignored.

FIFTY NINE

"Something Fishy"

A company called Aqua Advantage wanted to market salmon, trout, flounder and tilapia modified to grow from egg size to market size twice as fast as the old-fashioned way. They had growers all over the world, and they promoted the fish as the healthiest thing you could eat.

The biggest push was to develop crops that were not affected by pesticides, which meant that insects and weeds could be controlled without hurting the crops. It was a farmer's dream, and it seemed to be coming true.

Still, representatives of the Environmental Defense Fund noted that altered food could cause allergies to build up, and of course, the super weeds. It was a lot like in the late 20th century when antibiotics were overused and they stopped being effective because bacteria had evolved to the point they were resistant to the antibiotics. The truth was, no one knew what would result from these engineering "marvels."

Congress more or less agreed, but did nothing except suggest that researchers must prevent pest-resistant genes from spreading to weeds, if they could.

Representatives of the Center for Food Safety asked for a moratorium on distributing the "new foods," until more research could be done.

SIXTY

You Say Tomato

enator Orland Ingersol sat at his desk contemplating suicide. Not real suicide. Political suicide. The politically radioactive bill sat on his desk with his staff's scribblings and recommendations. He could follow his conscience and vote yes, but was it good for the country? And if he did vote yes, he would be put under a high-powered media microscope. His constituents would question his motives. On the other hand, if it ended his career, his purposes would be realized when people saw the millions of dollars flowing into his pocket. Ah, the life of a Senator in 2038.

The thought of this hot potato bill was erased when Clara entered the office.

"Senator? Mr. Dodson is here." Good, a respite.

"Well, let's not keep him waiting, Clara."

The middle-aged woman shared a conspiratorial smile with the Senator.

"Yes, sir."

Ingersol closed the file and quickly stuffed it in the lower drawer of the desk. No sense in letting wandering eyes catch a glimpse of political suicide.

Dale Dodson was one of Senator Ingersol's biggest contributors. The fifty year old ex-Marine, sometimes gentleman farmer and current lobbyist had poured millions into the Senator's re-election coffers.

Ingersol was a silent partner in Dodson's corporation called, "Green Grow the Lilacs." Dodson's farm consolidations did not exactly grow lilacs in southern Oklahoma. For three years, they had been experimenting with all sorts of seed hybrids and the results spoke for themselves. Fifty pound watermelons, grapes the size of tennis balls and pumpkins bigger than a little red wagon could hold. They even had trouble fitting into the trunk of a Volkswagen Beetle.

Dodson's agri-scientists had developed a greenhouse that was compatible with the new seeds, implementing solar panels and radiation treatment. Tropical fruits were being grown in Middle American during winter. Rich, lush papaya and mango that tasted sweeter than anything found in Hawaii or the tropics were being developed.

About every two months, Dodson brought Orland samples from the latest crops. He was not disappointed when the man entered the office with a potato sack.

"Potatoes, Dodson?" he asked with a gleam in his eye.

The old ex-Marine, weathered from the outdoors and still quite brawny, hefted the sack up on a chair next to the desk.

"You say potato, I say tomato," Dodson said in a sing-song. He reached into the sack and took out a juicy looking tomato that was the size of a softball. "Fresh from our newest hothouse."

The rich color had a sheen that beckoned the taste buds. Senator Orland Ingersol had a weakness for tomatoes.

"Be still my heart," he said. Then, he waved at the intercom. "Clara, come in here for a minute."

Clara entered with a look of anticipation.

Before he could speak, she started.

"I saw them when Mr. Dodson brought them in."

"Give Clara a couple, Dale."

Dodson handed the secretary two luscious specimens.

"Thank you, Mr. Dodson." Orland gestured towards the outer office.

The Senator smiled. "Set up a buffet for the staff, Clara. We'll dine in today on BLT's."

"Yes, sir." She fought the urge to salute and light stepped back into her office.

Dale Dodson handed one to Ingersol. "Here you go, Senator."

Ingersol hefted it. "It's heavy."

"Heavy with flavor. Bursting with taste."

"You sound like our ad group, Dale."

"That's the line we're using in the campaign."

Orland Ingersol held the softball-sized vegetable up to the light. "I can almost see my reflection in the skin. You want to stay for lunch?"

Dodson patted his stomach.

"I've got to save space for the chicken dinner I'm having with Senator Spiller."

"You're going to miss out."

Actually, Dale Dodson had had his share of the tomatoes. Good as they were, he had burned out on them after numerous taste tests.

"I'll let you have all the fun, Senator."

Orland laughed. "Okay, go grow me some green beans. I love green beans."

"It's already in the works, Orland."

After Dale Dodson left, Orland Ingersol took out his knife. He spread the sports section across his desktop and placed the tomato in the center. "Let's take a quick test drive on this baby." It took some effort to slice through the dense, red orb. Fresh juice spurted out. He felt a seed land on his left cheek. "Juicy, very juicy." He quartered the tomato and popped a piece in his mouth. This was a full mouth's worth, as he chewed, breaking down the contents. "Mmm." He moaned in silent delight. As he finished the first quarter slice, he savored the juices and held it in his mouth a while longer before swallowing. Orland smacked his lips. "That is one good tomato, like my grandmamma used to grow. No, she never grew them like this. They wouldn't last as long as these do."

He cut another quarter in half so the food was more manageable. The tangy taste was smoother than a regular tomato. He looked out his window at the Washington Monument, stark against a blue sky.

"This is what being a Senator is all about, perks." It did occur to him that his bank account had grown along with the tomatoes.

His wrist suddenly began to itch. He gave it a rub, but the unpleasant sensation did not go away. He started to scratch. Suddenly, a ridge of red bumps appeared on his arm. Hives? Soon, he was itching all over. His face felt hot.

"Clara?" Then he felt like a weight was on his chest. Orland was having trouble breathing. "Clara!"

Senator Ingersol was rushed to the emergency room, his tearful secretary by his side. When the doctor saw him, Orland's face was red and he was going into cardiac arrest.

"I need the cart! Stat!" A medical team materialized from down the hall. Orland's eyes were wide as his body convulsed. Someone was shouting at Clara.

"Has he eaten anything?!" It was the doctor.

She had seen the half-eaten tomato on his desk.

"He was eating a tomato, but..." Clara knew that the Senator was allergic to peanuts. As a boy, he experienced a similar episode when his brother and he had a bag of peanuts at a baseball game. Dale Dodson did not know of the deadly allergy. If so, he would never have presented the tomato to the Senator. "Green Grow the Lilacs" had been using a soybean blend with a peanut extract to engineer this crop of tomatoes.

Sixty One

Another Chance

Senator Ingersol recovered from his near death experience, although he could not convince the doctors that the tomatoes were the culprits. They insisted on peanuts which he had not eaten to his knowledge. All of a sudden the large melons and fantastic grapes did not hold the allure as before. When he was up to it, he summoned Dale Dodson to his office.

Dale had visited the Senator in the hospital and was deeply concerned. At this point, Orland wanted to know more about the operation in which he was a silent partner.

"Why did you use the peanut extract, Dale?"

"It was cheap and suited our purposes. We were piggy backing on what they were doing at the University of Illinois."

"Genetic foods."

"Just like we do at "Green Grow the Lilacs." We're developing tomatoes that produce lycopene."

"Lyco-what?"

"Lycopene. It's an antioxidant that might reduce cancer risk."

That's a benefit, thought Orland. Or it sounded like one at least.

Dale continued. "We're engineering beans that have few carbohydrates that produce less methane gas."

The Senator laughed but then arranged his face in a serious expression.

"Orland, we're using several exciting hybrids to produce broccoli with more cancer-fighting chemicals than conventional varieties. We have a culture group in soybeans with high levels of isoflavones. You know what that means? It can lower heart disease and risk of heart attack," Dale said.

Despite his misgivings, Senator Ingersol knew that Dale's research was noble. His friend had been truly distressed over the tomato incident. He decided a little levity was in order.

"Lower risk of heart attack, huh?"

"Exactly."

Orland broke into a smile. "So I could enjoy one of your peanut based tomatoes as long as I have soybeans on the side."

Dale didn't take the joke. "Orland, I'm serious. The engineering we're doing will save billions of dollars in the health care industry."

Orland sobered. He was thinking about the environmental groups that had been raising hell about the "new foods." They were demanding that more research be done before the foods were offered to the public. These were not wild-eyed tree huggers. To Orland, their reasoning made sense. He didn't realize that Dale was still talking.

"...potatoes with high starch content. We do that and they absorb less oil during frying, thus giving you a healthier, yet tasty French fry."

Orland held up his hand. "Dale, you don't have to sell me. We just need to make sure the contents in this stuff is disclosed."

"Once we're ready to package, we'll have everything listed, I promise."

Senator Ingersol stood up.

"All of this food talk is making me hungry. Let's hit the Congress Club. My treat."

Not long after, "Green Grow the Lilacs" went online, producing genetic foods for three grocery chains. The products were rubber-stamped by the FDA and when concern was raised from various quarters of the populace, a congressional committee was set up to investigate the new hybrids. Every few weeks, a news release would be issued announcing "The committee is not aware of any evidence suggesting that foods on the market today are unsafe to eat as a result of genetic modification." Then they added, "We found no distinction between the health and environmental risks posed by plants modified through modern genetic engineering techniques." Then the public information officer would smile, do an about- face and disappear into the silent gray building.

No one investigated "Green Grow the Lilacs" for their political contributions to members of the committee. At some point, Congress held hearings and again, the concerns for further research was to be addressed. Senator Ingersol was still an interested party, but was spending more time on his Oklahoma ranch as he was feeling poorly. Within seven months more or less, a committee was formed to appoint a research team to nail down any possible defects in the new food.

A year and a half later, the research team picked out its office furnishings and argued about which computers to buy. Two years after its initiation, the committee held its first hearing. Two of the Congressmen who had sponsored the committee were at death's door. The reason wasn't known, but they seemed simply to have worn out.

Sixty Two

Foreign Affairs

In Europe and Asia, the government was more cautious. And snack food maker Frito-Lay Inc., headquartered in Dallas, Texas, did not use the modified crops at all.

Nothing could stop the use of the altered food crop. It was like a tsunami of corn and wheat and tomatoes and everything else that mankind ate.

Just after his time at the Naval Academy in Annapolis, Jake had met a girl, another cadet and without thinking anything through, they had married, conceived a child, both been shipped to different locations, and quickly ended the ill-advised union. Except that there was a little girl they shuffled back and forth between themselves as well as to their parents. Her name was Eliza.

Sixty Three

Last Hope

Near dusk, a weary Colby rested beside the road, sipping from her canteen. Iona spoke to her like a camp counselor. "Time to rest."

"That's what I'm doing," Colby said irritably.

Iona spoke in an injured voice. "I am performing my assigned task."

Colby relented. Sometimes it was as if Iona were a puppy without the fur. "Sorry. I have to talk to the Commandant. And then we have to get back."

For a moment, she seemed to sag, staring across a great expanse of countryside, and in the distance saw Rice, Texas. But swiftly, she plugged earphones into Iona and pressed a button.

"Come in, Commander Addington. Earth to Code G. Earth to..." Colby listened, but her eyes were on the stars as she searched for the one thing that seemed certain, the space station. But where was he? Where was the one man who seemed able to help her? She shivered against the cold, and Iona was uncharacteristically silent, searching for the transmission.

Sixty Four

Unheeded Warnings

Before, when Colby had time, she finished the seemingly endless stack of reports her father had gathered. It was amazing how myopic the American media was. They lived a day-to-day existence, and rarely saw patterns in anything. It was the staid British House of Lords that sounded the alarm.

Interestingly enough, in Colby's view, many of the crustiest leaders were old guys who were so-called country gentlemen, much as had been the late King Charles. Throughout his long life, he had been interested in all things agricultural, as were many in the ruling class. Even though they sensed the danger of the altered foods, it took a long time for him and his colleagues to be heard. By then, many agribusinesses were so far down the road to genetic manipulation, that for most it was too late.

As early as 2000, more than forty years earlier, there was talk of tightening food rules for genetically altered foods, but the U.S. Congress had extensive ties with agribusiness, particularly in the Mississippi and Arkansas delta regions. The farmers there were wealthy, growing rice, cotton and sometimes nothing at all. The completion of research in 1999 of a strain of bioengineered rice and the subsequent declaration by the FDA and the USDA announcing that there were no health or environmental concerns was cause for celebration, not questioning.

Further, a USDA spokesman let it be known that the rice was as safe as rice grown in the traditional way. No longer would there be any need for federal oversight of the growers. They were let loose, and the world media did not take note.

In 2009, there was an international conference in California, attended by more than 15,000 scientists, chefs and businessmen whose focus was the "new" food, which meant genetically altered. The irony of the popular meeting was that the United States Department of Agriculture had already decided that the new foods were a boon to everyone. They had put their stamp of approval on it, as had the Food and Drug Administration.

Floating around the edges of the gathering were reports that genetically altered foods and the procedures that made that possible might lead to cancer. No one wanted to hear about it.

Certainly no one dreamed that there was something worse than a malignancy, and that was the progeria that did result from the scientists' work. In London, the House of Lords assigned the nation's Center for Food Safety to investigate the use of the nano materials, which seemed to be used to make products look better or pour better. There was no publicity about the rapidly increasing usage. Most people had no idea. When the potential destructiveness of the policy became known, it was too late. One thing that did happen was that there was no more talk of altered foods in advertising, sales or from government. The attitude was let the potato chips fall where they may. In Europe, however, people had been made aware and they did think they had a choice not to buy the "new" food.

In the year 2000, a small company recalled some taco shells. The story was that the shells were made from genetically engineered corn which was mistakenly used. It had earlier been approved for animal consumption, but "accidentally" got into the human food chain. Again Congress was heard from. They had nothing in place throughout the federal government that required mandatory safety tests. And despite early efforts by U.S. Rep. Dennis Kucinich, an Ohio Democrat, who sponsored legislation to require labeling and safety testing of such foods, they never would put anything significant in place that would work.

SIXTY FIVE

Her Voice

Addington stared at his receiver. She had not answered on the first or second transmissions. Despite a week of trying, he had given up on her. He had assumed as had his command staff that Colby had caught the dread disease decimating earth, progeria was the named diagnosis, although it could be some mutation of that. Probably she had simply gone somewhere to die.

"Jake, Commander Addington! Come in please," spoke the voice of Iona. Then another voice picked up. "Jake. Answer me!" Her voice sounded loud and clear.

"Colby! Thank God!"

Colby laughed. "Colonel, sorry. I haven't been able to contact you when we planned. I've been on this crazy search. Whatever the Life Source is, I don't have it. Is there anything...have you found out anything?"

Addington hesitated. "Maybe. The reversals in the aging process are still happening, but not enough for us to act on. We did receive a message from Norway. Looks like they anticipated something like this. I'll get it to you."

"Can you transport food, water?

Addington nodded. "In about three days. We can get it within a couple of miles of the town. Can you be ready to receive it?"

"I can pick it up. Will it buy us time?"

Addington spoke slowly. "Some time. Yes, but those men with you. Do you know anything more about them?"

"Not much. One was in service in the Antarctica. The other...who knows?"

"Is this plague on the wane? I can't tell anything from here," she said.

Jake hesitated. "Maybe," he said, but his words were tentative.

"Meaning you don't know."

"Right," he agreed. "The battle lines were drawn a long time ago. What we do know is that this is a battle of seeds, and the war started in the 1990s. The consumer didn't know what was happening, and the journalists didn't pick up on it. They were too busy pouncing on trivia. Anyway, the pro hybrids started it and it worked at first, but they took it too far, and we got this disaster. I'll tell you more later, Colby. Take care," Jake signed off.

"Right." The transmission ended, leaving Colby feeling alone. What was it about Jake that always left her wondering what else he knew. She unplugged her earphones and stared at Iona.

"At lease he knows we're alive," she said.

Iona's soft whir was her only response.

Soon Colby was out on the highway, aiming for Rice and her mysterious companions. Maybe she could find out more about them while she was waiting for the supply drop.

Inside the hay filled railroad car, Angel touched his rifle gently every few moments, as if he were in a deer blind. He was confident she would be along soon.

———

Jake did what he promised quickly. He texted her additional information, using high speed radio transmissions that were encrypted. When she had finished reading the material, she knew pretty much everything about the plague. She also knew that there was an answer, an antidote, somewhere in the county and that it was up to her to find it. But could she trust Sam? And what of Potter?

SIXTY SIX

Drowning In A Watershed Moment

There was a watershed moment in 2006 when Arkansas's more than 4,000 rice farmers filed a class action lawsuit against Riceland Foods who were headquartered in Stuttgart, Arkansas, once the duck capital of the world. It was a place where nature was revered as a positive thing. The accusation was that Riceland exposed the rice farmers to an "ultra-hazardous risk." What the giant firm had done was experiment with genetically modified rice which got loose into the food chain and contaminated the commercial supply of rice. The unfortunate event or series of events led to the European Union refusing to import Arkansas rice. The strain was called "Liberty Link," and it contaminated everything it came in contact with, according to the law suit.

Some leading scientists warned of what would happen. They knew that there was a risk of contamination, and their analysis was correct. Somehow, the "Franken food" rice found its way into research seed production fields and spread across the state's rice producing farm lands in the Arkansas and Mississippi Delta, highly productive farms that fed millions. And while the United States Department of Agriculture had not officially opined that the rice was harmful to humans, it did not approve the rice to be consumed by humans either. The Europeans, more careful about their food than the U.S., discovered the genetically modified rice in what had recently become routine testing. They refused to accept that shipment or any other from the United States. They knew that Pandora's box was open, but they had no idea what troubles were to come.

As late as 2010, the case had not gone to court. The genetically engineered rice was not contained. Ironically, the company was a farmer-owned cooperative and the world's largest rice miller and marketer. So it was farmer co-op versus individual farmer. Still, farmers began planting the original rice seed again, more than a million and a half acres of it. The question on everyone's mind, was it too late?

There were other events, hardly noticed, there was so much noise from the media, which prattled endlessly usually about nothing. So when the "salmon uproar" occurred, few paid attention. A virus had threatened farmed salmon, bioengineered fish, causing a general collapse. That might've been for the good, sent restaurants and fisheries back to natural salmon, but the demand was too great. The food engineers found a way to ignore the virus, and farmed fish became an even bigger industry. The story was soon forgotten.

Meanwhile, during the rice uproar, another story emerged from North Dakota. Scientists had been experimenting with canola, a variety of rapeseed which is of the mustard family. In its natural state canola contains reduced levels of erucic acid, making its oil palatable for human consumption. The second element was a reduced level of atoxic glucosin, which made the meal desirable as livestock feed for which it originally had been used, hogs, sheep and cattle.

The canola had been altered, and the 'new' product had been found growing alongside roads in North Dakota, among the first times that a biotech crop was found growing in the wild. The event had been predicted by scientists at the University of Arkansas. A young woman named Meredith G. Schafer, a graduate student, had traveled more than 3,000 miles of North Dakota roads, from interstate down to the smallest county road, stopping every five miles to take samples of any canola plants her team spotted.

Altogether they collected more than six hundred plants and discovered that 80 per cent of them were genetically engineered. It was an 'aha' moment, and a frightening one. The canola had a resistance to

the herbicides available, which meant that they had become uncontrol-
lable weeds. So much for the myth that they would not propagate. They
could and did. The scientists shared their information, and came to the
ultimate conclusion that they would deal with the strain of weed with
mowing, and by developing different, more effective herbicides.

Sixty Seven

Farmer Of The Year

He broke the record for corn production, the sixth year after he had begun using the new hybrid seed. One acre produced 394 bushels. Present to see the amazing feat were USDA officials, state agriculture agents, dozens of other farmers and of course, newspaper and television crews. Franklin Compton would eventually be named national Farmer of the Year, a temporary honor as it turned out. But Manchester, Iowa, was the talk of the nation, for a few minutes on the morning news at least.

Compton's methods also included chemical fertilizers and of course pesticides. But he credited the hybrid seeds for tripling his corn output. His goal was to pass the four hundred bushels per acre mark the next year. He hardly gave a thought to any warnings, that environmentalists were worried about, that there might be unintended consequences in the use of the modified genes. The crazy thing was that corn was the easiest thing in the world to grow, and the year Compton made his breakthrough, corn prices had dropped abnormally low, simply because there was so much corn available. His breakthrough was not particularly needed. It was much like the cloning of cats done at Texas A&M University years before. There was, as most people knew, an ample cat population. At the same university a scientist had figured out how to turn cottonseed into food. It was a quest that had lasted for decades. It was the answer to prayer, for it could help feed millions from an easily grown crop.

Still, it was exciting to set a record and be recognized for it. There had been a time when farmers were all but forgotten, but lately, the incredible advances in crop production had brought them back to the forefront. The honor would be recognized nationwide. It was gratifying indeed.

Sixty Eight

School Daze

Inside the school, the late afternoon sun sent streaks of gold through the clerestory above. Tiny particles rode the light like sea anemones on a high tide.

Sam stared into the trophy case, first at his own trophies, and then at the silver canister behind his photograph. Somehow he was not expecting that, and it took a moment for him to register what it was. When he did, he began to run back and forth along the front of the glass enclosed case trying to find a way in. What the hell, he thought, picked up a school desk chair and heaved it through the case. The glass shattered in all directions, but he had his arms over his face to protect himself. Once the glass and dust settled, he pulled out the canister. He was pretty sure he had found the time capsule, the source of life, according to Colby. If only she were right, he thought.

Sam stared at the container and turned it over and over in his hands. He expected to find a button or a twist-off cap, but there was nothing apparent. After a while, Sam decided he needed tools to get the canister open.

He wondered if Potter might know where tools were. He seemed to know a lot about the place, where people lived, some of what had happened. How hard would it be to find a screwdriver?

Sam started out across the football field, carrying the newly found treasure or it would be if he could get it open.

Out on the field, Potter kicked at the football, missed, and almost fell. So he picked up the ball, and kicked at it again, his effort still futile. Then he saw Sam walking toward him. He grinned and rushed to meet him.

"Wanna play, Sam? Now?" he pressed.

Sam didn't stop. "I need some help," he said.

Potter spotted the canister and chased Sam like a kid warting his older brother. "What's in that? Could you open it? Maybe it's candy bars?"

Sam spoke as if to a wayward child. "Maybe. Probably not."

Potter pleaded, "Please, please."

"I have to figure out how. I'm gonna need some tools. Will you give me a minute?" Sam responded.

Potter abruptly stopped nagging, as if he were remembering a lesson from his childhood. "OK."

Sixty Nine

Follow The Sun

A subdued orange sun was still high on the horizon as Colby raced toward the town. Near the unmown hay field, the four wheeler without warning abruptly stopped in its tracks. Irritated, Colby climbed off and began to push it, talking to herself, or to Iona which was her habit. It was probably a failure to get enough sun.

"They've got the cylinder. Sam found it. He sent me a message, I don't know why."

Iona's voice was conciliatory. "I'm sure Sam didn't know, Colby. He would share with you like he has now. He called and told you."

Colby was not moved. Her voice was angry. "I'm sure Sam did know, and I'm just as sure he won't share what's in it with me."

"It grieves me to think so," Iona murmured.

"You're a robot for heaven's sake. You don't have to grieve. It was my information that led him to it," she said, still irritable.

She pushed harder as they began ascending a hill. Out in the field, hidden behind a low stack of hay bales, Angel steadied his rifle and adjusted his scope. He then took aim out toward the highway.

In the town, Sam worked at the canister like a demon, trying everything he could think of to get it open, twisting and turning it, studying it, hammering on it finally, all to no avail. Potter sat patiently nearby, watching with interest.

Sam pitched the canister out in front of him, disgusted. "There's some kind of formula, some key to open the thing. Colby has it. She'll be back soon."

"When?" Potter asked, his eyes benign.

"Soon," Sam offered.

Potter smiled. "If there's candy somewhere,..."

Sam shook his head, his expression a little puzzled. He gave the big guy the candy from his pocket. "It's all yours." The last thing on his mind was candy, and why would this bozo be worrying about something that trivial anyway? Surely he had more to consider, but he did too and allowed the thought to drift away.

SEVENTY

Answers

On the roadway, Colby had at last started the ATV again. She revved the throttle and began to pick up speed.

From his hidden place behind the bale of hay, Angel wondered if he could hit her. His increasingly shaking hands made it difficult, but he saw her coming over the hill and instinctively pulled the trigger. The shot echoed through the woods nearby.

Colby heard the first shot and reacted swiftly, ducking her head and trying to see behind her. The cycle swerved, but she righted it. Startled, she looked back again, but she could see nothing except a line of rotting hay bales in the field. She leaned over the handle bars to give a low profile to the shooter and headed toward the town. The shooter fired twice more.

She felt something tear at her, something hot, searing, as if a piece of molten metal cut through her arm. She held tightly to the four-wheeler's bars, but then her grip loosened and she turned the bars loose. She could not feel her hand and arm. The four wheeler tore out of her control and careened into a ditch.

On the main street, Sam heard the shots reverberating against the deteriorating buildings. He stood perfectly still for an instant, listening to the reflected sound around him. His first thought was the motorcycle, but it would not start. He grabbed the key in case Potter might have better luck, but then he was running full out toward the highway. It was Angel or Dave, he knew it. But had they hit their mark? He prayed that they had not.

Potter was slower to react, in fact, he watched Sam race away, and instead of following as he usually did, picked up the cylinder and began trying to force it open.

From the field, Angel and Dave tried to hurry toward Colby to see what they'd done. Chuck followed behind but their movement was as if it were in slow motion. Their time was almost finished.

———•———

Sam stopped once for breath at the edge of town, and then he continued to run out along the blacktop.

In the middle of the football field, Potter turned the cylinder every direction he could think of, and then, like Sam, tossed it onto the ground. He got to his feet, and walked over to the motorcycle. He climbed on it, and kicked it once or twice, as if it were a horse. He got off the machine, and laboriously began to push it. He did not notice that the key in the ignition was gone.

———•———

The three men stood around their prey. Dave leaned down and struggled to pull Colby's helmet off, but it was buttoned on tightly and he gave up.

"She's dead," he said.

The other two, Angel and Chuck, nodded wordlessly, their sad faces speaking volumes. Written in the deeply etched lines were envy, curiosity, despair, questioning and hopelessness. They turned and walked away.

———

Sam, out of breath from running, dropped to his knees, and sucked air into his burning lungs. Then he started again, running along the road.

Colby waited until she knew they were gone, and then crawled toward the four-wheeler and pulled Iona from her backpack. She tried to turn the automaton on, but couldn't. Is this what it feels like to die, she wondered? Even a simple task is too much?

"Gimme a minute. Just a minute," she whispered, to no one in particular.

At the edge of town, a grim-faced Potter pushed the motorcycle, stopping every now and then to stare at it in confusion. He saw them crossing the field, rifles on their shoulders, Angel and Dave. Chuck had stopped to rest a few yards back. They returned his wave, but without joy.

When he reached them pushing the motorcycle, Chuck looked the machine over and pointed to the keyless ignition. He glanced at Angel who rolled his eyes. Potter looked surprised. Angel took his arm and whispered to him that they had killed Colby. Potter sagged at the news, his eyes tearing up. He started to where they pointed, but Angel took his arm and they turned back to the town.

A half mile away, Sam tripped, but he picked himself up and examined his hands. The thought nagged at him. Was he aging too? He went on.

When he got to her in the growing dusk, Colby was resting on her knapsack. She attempted to grasp her canteen, but couldn't manage it. Gently, Sam gave her a drink, and then carefully uprighted the four-wheeler and got her aboard. He secured her with his belt and started off slowly.

———

Sam managed to get her back to camp. Later, Sam sat beside Colby as she rested on the cot. He'd take her to her parent's home later. Something had changed in her attitude toward the man who had rescued her.

"You have to meet the supply drop for me," she said.

"I'll be there on time but I don't want to leave you. They're still out there."

"You have to. It'll take you...maybe two hours, a little more."

"I know. I know." He glanced at his watch. Funny, how despite everything, they still followed the hours.

"We don't have enough time if you don't go now."

"I'll get there."

He handed her the closed cylinder he'd picked up from the field. "Keep this safe, until I get back. There's some simple combination, something."

"Maybe I'll solve the puzzle. I was always good at puzzles," she said, attempting to lighten the moment.

"You just might. I don't know where Potter will go. He has a long hike back here. Take this."

Sam pressed a gun into her hands. Quickly she tucked it under her blanket.

"Bye." He hesitated near the entrance to the tent.

"So go," she said.

He grabbed his backpack and started through the flap, but then he dropped the backpack off his shoulders and he turned back to her. In a moment, he was kissing her and she was returning the kiss.

He took the ATV and was flying over the road as fast as the four-wheeler would go. His thoughts were to complete the mission of meeting the supplies and then return to her. After that was done they would decide together.

He didn't like the idea of leaving her alone, not for a minute, but he had no choice. Besides, she was a feisty girl, and no doubt could handle herself in a one on one situation.

Inside the tent, Colby listened to quiet music and then dozed off. The round of antibiotics and painkillers soon allowed her a deep sleep.

———•———

Potter was hungry. The Old Ones had gone and left him alone. There was one place he knew there was food. Shortly, he stood outside for a minute, trying to decide what to do. Where was Sam? Where was the ATV? In his mind, Colby was out beside the road. The coast was clear. He slipped quietly into the tent. Surprised, he looked down at Colby, a gentle expression on his face, but something in the corner attracted his attention, the food supply. Potter reached into the box and frantically pulled out packets of food. With his teeth, he tore into the packets and gobbled the food like a dog left to its own devices. When he had finished eating, he curled up on the floor at the end of her cot and slept.

SEVENTY ONE

The Future

Colby awoke and found Potter still asleep. She could see that he had been robbing the cache of food. Feeling the pain and stiffness of her wound, she slipped out of the tent. This is what it's like to be old, she thought. She was waiting for Potter when he emerged sleepy eyed with a fistful of health bars in his hand.

"Coffee?" she offered, but then remembered that he loved hot chocolate with marshmallows. She explained carefully how to make it. In a little while, he grinned up at her and began to sip the hot drink.

She studied his features. Something was different. His eyebrows? Were they whiter? His brow. Was it more wrinkled? She wasn't sure.

"I want you to tell me about your people, what happened here, everything," she said.

Potter licked the chocolate off his upper lip and began. "We didn't know how serious it was. Crazy things were happening to us, to our bodies."

Colby began to register what he was saying. Her voice was gentle. She didn't want to rattle him. "Who is 'we,' Potter?"

"Me and my friends at school mainly."

Colby stared at the man. "What happened to your friends?"

Potter frowned. "There weren't that many to start with. They always had about enough to field an eleven-man football team."

"You played football?" she asked.

"Oh, no ma'am. You don't play football 'til high school."

Potter smiled as if he'd explained everything.

"High school? Eh?" For a moment she appeared confused. "How old are you?"

Potter thought it over. "I was born in...2028."

"You were born...fourteen years ago?"

"Yes, in September I'll be fifteen. I grew this tall...in a month. My daddy is still out at the farm. He couldn't do anything to stop this. Can you help me?"

"I'll try."

—◆—

Sam waited at the airfield, frequently glancing skyward, and growing increasingly discouraged. What if they couldn't make the drop? What if the payload vehicle couldn't blast its way through the atmosphere? What if it simply burned up? The questions were ones he'd learned to ask over years of watching space exploration. His discouragement was growing. The sky had never seemed this vast and empty. He saw the dark specks first. They were gaining in size, balloons that were settling downward, propelled toward the airfield carrying their payload. Jake had come through.

He drove toward the materials as they landed. It had been their last hope, and it was at hand. Wafting to the earth were huge boxes of supplies that set down with a "whummp."

Sam offered a snappy salute, but he knew the crew managing the operation could not see him. They were high above the earth, driving the game-like controls from Polaris.

The first thing he did was search for the information container. He found it, and popped it open. It was reams of instructions about horticulture, farming, animal husbandry and anything else a young well-educated farmer would need.

Sam nodded his head without much joy. Here were their answers. Soon they must get in touch with Norway. People would be waiting to hear from them thanks to Colonel Addington. This was the future and he meant to get it right.

Seventy Two

The Well

Inside the tent, Colby slept restlessly. She had instructed Potter to go out to the farm and check on his father, anything to get him to leave. He had gone, but reluctantly, this time walking, having given up on the heavy cycle.

A large hand brushed her face and she awoke. She struggled to get upright, her gun in her hand. She was breathing hard. Something was terribly wrong. She looked around the room. Iona and the cylinder were gone.

Colby armed herself with the pistol and set off for downtown. She hadn't gone far when she heard World War II music blaring down the empty street. It was Iona, the volume turned up to the maximum. "I'll be seeing you in all the old familiar places."

Potter set the little robot on the ledge of the well and began puttering with the cylinder. He thumped it and nothing happened. Colby hid behind a car and watched him hit it again against the ledge. Iona bounced closer to the well's wide opening.

"Duh, duh, don't," Iona pleaded, her automatic voice suddenly sounding alarmed.

Potter stared at the cylinder. Something came to him and he turned to the robot. He put the cylinder in front of Iona.

Iona studied the cylinder. A rapid array of numbers clicked across her screen, and then stopped. Nothing happened. "Sorry. Does not compute. Antiquated lock, before my creation. Sorry. "

Potter picked up Iona and shook her hard. "Try it again!!"

"Yes, sir." she said.

Again, the numbers whirred and beeped, but nothing happened.

Potter's face turned crimson with fury. He picked up the miniature robot and held her over the well. "I warned you!"

"I did try, Mr. Potter. I did my best."

"You get it right, or you go swimming!"

"I can't swim!"

———

Sam revved the ATV until it hummed like a new jet fighter taking off from an aircraft carrier. It skimmed along the highway. He'd left the stores behind. They would need to find a truck to haul all of the materials, but now, he had to get back!

———

On the main street, Potter held Iona over the well.

"Give me the numbers!" Potter yelled.

Terrified, Iona squeaked, "I have a chipped microdot!"

Potter shook her hard. "The numbers!" he screamed.

Inexplicably, Iona began to sing an "80s" hit, "Hit the road, Jack..."

Potter understood nothing and then he released her into the well. He heard a loud "ooooohhhhh," and then a splash.

"You are gone!!" he shouted.

Colby moved slowly down the main street, and soon saw Potter staring curiously down into the well.

Colby came up behind him and waited. He was too big for her to take on. But then the noise of the four-wheeler caused them both to turn their heads. Potter's eyes reflected a mindless fright as Sam leapt at the stupefied man child who threw up his arms to protect himself. The cylinder rolled a few feet away.

Colby left Potter to Sam and raced back to the well, frantically calling for Iona.

Sam dived at Potter's knees and heard the bigger man cry out. But then Potter kicked Sam hard, smacking his hand with his heavy military boot. Sam backed away and checked his hand. Potter grabbed awkwardly for the gun at his side, which Sam had forgotten. As he reached for it, Sam rammed his shoulder into Potter's gut. The gun went flying. The stouter Potter pushed Sam backward and raced toward the well. He grabbed Colby, who was leaning over the ledge looking for Iona. She pulled out her pistol, and hoping to scare Potter, she fired it down the street where it bounced off an old Exxon-Mobil sign.

Potter looked surprised, and released Colby as if she were a rattlesnake.

"It's real!"

Colby pointed it at Potter. "Yeah, it's real. What'd you think?"

Potter threw up his hands, and snuffling loudly said, "Don't shoot. I give up!"

Sam glanced at Colby. "Let me get something, anything, to tie him up."

Sam found a piece of rope and tied up Potter who did nothing but stare at Colby.

"Help me! Help me!" came from the well.

Sam and Colby leaned over the well and looked down. "Oh, my lord, is she drowning?"

"We've got to get her out of there."

Sam grabbed the first thing he saw, the water bucket. Quickly he lowered it into the well while Colby watched anxiously.

"Help!" The little voice echoed from deep within, like a child crying out. In a second, Sam pulled her to the surface, and Iona was staring over the edge of the bucket, freed from the well. "It's about time! How long does anybody expect me to float down there in the dark? Spiders and snakes and minerals that'll rust me! Dangerous!"

Colby lifted Iona from the bucket and began drying the little robot off.

Sam pulled Potter to his feet and held him. "We've got to lock him up."

"The old depot has doors you can lock, I think," Colby offered.

Sam quickly led Potter to the steps of the depot with Potter pulling against him all the way, shaking his head "no," but Sam had the leverage of the rope and managed to push him inside. He found a room with a door and locked the big man there. "That should hold him."

Colby began to realize that time was running out on them.

"We need to get into the cylinder," Colby said.

Sam looked interested. "Use dynamite."

Colby looked puzzled, "What?"

Sam looked less than serious.

"Rrrrrrrrrr," said Iona .

Colby and Sam stared at Iona . She was making a sound like a cat purring, but nothing else came out. Colby was exasperated with the robot.

"Come on. It's a combination of numbers. How hard can it be?"

Sam frowned. "Just run combinations!"

Colby shook her head. "That could take a month. Are you all right?" she asked Iona.

Iona began her litany of grievances. "Of course I'm not all right. I almost drowned. I may rust. A spider sat on my volume control!"

Colby wasn't buying it. "You need to dry out. Alone, in the tent. And you will start the number sequence."

The robot's voice was accusing. "You just want to be alone with Sam."

Sam went over and picked her up. "It's the other way around."

Iona got it then. "Oh, I see."

He put Iona into the tent. When he came out, he stood for a moment looking at Colby, and then he walked to her and took her into his arms. She did not resist.

Later, Sam began to open up to Colby. He told her about Eliza and it felt good. She was the past, and if they were so blessed, Colby was the future. It was OK for it to be that way, he knew. The best thing was they had supplies. He had already found a pickup truck and a forklift that ran on solar powered batteries. He figured that would give them the power they needed to move all the supplies Jake Addington had sent them.

Seventy Three

Keeping On

The following morning, Potter was still sleeping on a cot that Sam had brought him. Something woke him, the sound of feet shuffling outside. He got up from his bed and approached the locked cell door and then scurried over to a window. Potter's hair had turned completely white overnight.

Old hands fumbled with the lock. Angel and Dave, ancient as bristlecone pines, stood near the door, watching as Dave, wizened as if he'd been frostbitten, struggled with the lock.

Potter came to the door and stared out at them. "I knew you'd come," he said.

Angel spoke. "They have the life source?"

Potter averted his eyes, shamed. "Sam took it away from me, tell Papa he did, Uncle Dave."

Angel looked resigned. "Your papa's out here, in the street. We brought him on the golf cart. He's...almost gone. Come out and see him. Tell him goodbye."

Potter looked worriedly out toward the street, but once the door flew open, he hobbled out. The cart was there and inside an ancient man.

"Papa," Potters said, and climbed in beside him.

The old man touched the younger one's silver white hair. "It's too late for us. Whatever they've found won't help us." He stared at his son. "And you, my son, it won't help you."

Potter turned and caught his own reflection in the glass window behind him. "Maybe." He whispered.

"There are only a few of us left," said Angel.

"Sam hurt me, " Potter wailed.

The man in the cart patted him on the head. "He didn't mean to. You come with us, son," he said.

Potter's face was the picture of a 7-year-old denied candy. "Want my gun," he whined.

Angel found a gun in the cart and gave it to Potter. Potter cradled it, a grin on his face. They left the downtown street then, and retreated to the old man's farm. Angel and the others, with help from Potter, who still possessed some strength, took the old man inside and made him comfortable. Then they left him.

SEVENTY FOUR

The Cave

It was cold in the cavern, even in the crew quarters, but the team that worked there had been ordered to remain, no matter what. There could be no chance of their becoming contaminated, no chance at all.

The job assigned them was vital beyond all others. The mission was to save the native seeds, to put them into a frozen state, and to guard them with their lives. If they were taken out too soon, it would be disastrous for whoever was left alive on earth. If they waited too late, there would be no time for the seeds to work their magic.

Whoever had chosen the location for the ultimate seed bank, had made a wise decision. Few could find their way there, and once they did, their purpose must be clear. The young crew took its mission seriously. They had no other choice.

From the outside, one saw frost and gray steel and locked gates. Even the guardhouses were impenetrable. But it was also important to know that those would come who needed to take the seed, and that would have to happen. Deciding if they were the right people to do it was left to young officers. Their task was to search the earth for survivors, and then home in on some who could shepherd the remaining people to the other side of the blight.

SEVENTY FIVE

Revelation

Inside the snug tent, Colby awakened and looked around her, anxious. Sam was asleep nearby.

"We have to get that thing open," Colby said, loudly enough to wake him.

Sam was instantly awake and on his feet, pulling on a jacket, and then tying his shoes.

"I'll give Potter something to eat. I expect to see that thing open when I get back." Sam grabbed a handful of food packets and a canteen and strode away.

Iona began to sing. "Hit the trail...hit the track...hit the road..."

Colby shook her head, and held up a hand as if she did not want to be bothered by Iona's chatter. "Time is running out. Food is running out. My patience I left back in space. Can't you concentrate?"

Iona went on singing her song. "40 ways..."

Colby listened more closely and she suddenly she began to get it.

"Forty...forty...forty...4 - 0, 4-0, 4-0..."

"My, aren't you a clever little human," said Iona.

SEVENTY SIX

A Puzzle

Sam drew water from the blessedly uncontaminated well and poured it into his canteen. He secured the top, and headed for the depot. What greeted him was a door torn off its hinges. There was only one person capable of that. Certainly none of the old men.

At the camp, Colby was sitting in front of the cylinder, staring at it, preparing to try the new numbers. "I want Sam to be here when I open it."

"Colby?" Iona's voice seemed strained.

"Yes?" Colby was staring at the cylinder, but at that instant, a blurred figure launched himself on her from behind. The pair fell into the dirt and struggled. She pushed him off, surprised that she could do it, but instinctively she knew that he was weaker now. He reached for the cylinder and she tried to block him. Iona's warning siren screamed like some horror movie. Potter, startled, stopped fighting and looked at Iona. With that opportunity, Colby grabbed the cylinder and raced down the street.

Potter struggled to his feet and chased after her, fell over tangled feet, got up and went on.

On the main street, Colby stopped in front of the general store out of breath and trembling. She looked back over her shoulder and saw Potter lumbering toward her. She ducked inside.

———

Sam glanced up at the piles of boxes in the old warehouse. He heard a scraping sound and then saw the pile teetering, and they were falling on him. He tried to run, but the boxes tumbled into his path and held him there momentarily. He pushed them aside, and freed himself. He heard the sound again, like a large rat might make, and rushed toward it. He rounded the corner and saw the old man. For a moment, his eyes fixed on Sam, studying him, and then he reached out his hand. Sam went to him. He thought that he might know him, but he wasn't sure. "Who are you?"

The old man tried to speak, but he was exhausted.

"Who?" Sam asked again.

The old man looked at him with ancient eyes, tears welling in them. "I...taught you how to throw long." It was Chuck Ames. He was the former iron man, the school football coach, and he had fallen. But he could say no more. He was at the end of his strength. His eyes closed, and Sam knew that he was dead.

Inside the store, Colby still carried the cylinder and Iona. She looked for a place to hide. Outside, Potter was pulling at the door and screeching at her, a strange high-pitched sound that some animal might make.

The trapdoor inside the general store! She remembered her commando training, flipped up the hatch and jumped into the gaping hole. With one movement, she pulled the door shut.

Along the sidewalks, Potter lumbered like an aging elephant. At that moment, he caught an unexpected glimpse of his own reflection and stopped, staring at himself, touching his lined face, disbelieving. But then he pulled opened the door to the store and went inside.

A few blocks away, Sam left the warehouse and ran toward the main street, looking for Colby and Potter.

Colby held Iona close and made herself as small as possible. They waited in silence in their hiding place which was dark and dry. Colby lifted the trap door slightly and held Iona up for a peek.

Iona saw Potter's heavy military boots and then moved up his thick body. It was then she saw what he held in his hand, a large ax with a rusted edge. He was holding it like a batter about to bunt.

"That's far enough!" said Iona, whispering. Then, "Take me down!"

With a swift move, Colby pulled Iona quickly back into the darkness, out of Potter's sight.

Inside the store, he lurched from cabinet to cabinet, throwing open the doors, his eyes sweeping the empty expanses. He stared intently at the floor, and he spotted Colby's footprint. He grinned at his discovery and followed it back to the iron pull on the trapdoor. "I know where you are," he said in a playful singsong voice. He reached down and pulled at the door.

Under the floor, Colby poised to strike out, hoping to catch Potter off guard as he raised the trap door. She could think of nothing else, since there was no place to run.

Sam burst through the door, and rocketed into Potter, landing a hard jab squarely into his kidney. The big man fell to his knees and Sam kicked him between the shoulder blades with the sole of his boot. As Potter lay sprawled on the floor, Sam finished the encounter with a boot to the side of the head.

Dust sprinkled down on top of her. She waited a few seconds, and then pushed up the door. She saw Sam dragging Potter out to the street. She followed him and looked down at the beaten man, feeling some sort of compassion for him, even though he had attempted to destroy them.

Potter sat up. Every line in Potter's face was etched downward. He was pitiful as a lost puppy.

"Please give me the capsule. That's all I want and I'll go away, far away."

"Where do you think you'd go?" Sam asked, incredulous.

"To anywhere. New York maybe?" Potter said, looking one way and another, like a trapped animal.

Sam shook his head. "I doubt you know how to get to the highway. You need to go back to the warehouse. One of your buddies is dead."

"You killed him!" Potter accused.

"Heart failure, looks like," Sam said, not responding to the accusation.

But then Potter fumbled at his belt, and pulled out the pistol the old men gave him.

"Give it to me," Sam said. Potter looked as if he couldn't decide, but then tossed it to Sam.

"He must like us. He's never used that gun on us," Sam said.

Sam pulled back the hammer and aimed the gun at Colby, who was laughing at him, but then she saw Sam's finger squeeze the trigger and she stopped laughing. The gun fired with a staccato "pop, pop, pop." As Sam repeatedly pulled the trigger.

"A cap pistol?" she said.

Sam turned his attention back to Potter. "Get up. You're getting out of here."

Potter lay still, defeated. Sam started to pick him up, but hesitated. He had to find the life source. Potter wasn't going anywhere. He would come back for him later.

Outside, Colby and Sam plugged in the numbers to the cylinder, hoping that they got it right this time. They waited. Like some ancient crypt, there was a scraping sound, and then the cylinder opened. Colby and Sam peered inside. There were the usual items buried in a time capsule. This one held a Corsicana Daily Sun, a Military Might soldier, the signature of the then Governor of the State, a United States flag with fifty stars for the states, an Apple phone no bigger than your palm, a number of popular CDs, and a map.

Sam pulled out the simple map, and studied it. "It leads to..." he began.

She looked at him knowingly and completed his sentence. "The school."

SEVENTY SEVEN

Back To School

The vehicle took them there in minutes, and they were racing along the corridor, passing the trophy case, since they knew it was not there, looking into classrooms, pushing open doors, and at last they reached the biology laboratory. They hesitated, together shoved through the door, and stood in the middle of the lab and studied the room.

"It's under our noses. The life source is here. Right here," she said.

Sam began to circle the room as if he were stalking prey, but one that could cause fatal damage.

Colby searched in drawers, on shelves, anywhere she could think to look. Preoccupied with her search, she did not see the broad shadow at the door.

Potter crashed into the room and grabbed Colby, jerking her club from her belt and holding it against her throat.

"I can kill her with this," he roared at Sam.

"No. Don't."

"Give me the life source!" Potter demanded, his voice frantic, eerily breaking, sounding both like a teenager at puberty, and like an old man at the end of his life.

Sam frantically scanned the room. "I don't have it."

"What is it? Where is it?" Potter demanded.

Sam moved toward the pair.

"No, Sam!" Colby shouted.

Potter had a strange expression on his face, as if he were in pain. "Back off!" he shouted.

Sam raised his hands as if to placate Potter and stepped back.

Potter looked at Sam imploringly. He was clutching at his shirt, tearing it, and then grasping his chest and his throat as if he could not breath. He dropped the club, and then he fell to his knees beside Colby who was frozen, uncertain whether to run or stay. She looked at him and knew that something was terribly wrong.

She backed away from Potter as he struggled for breath. Then he pitched forward, his hand reaching out for something, and he lay still in the dust, this time for good.

Colby turned back toward the man, and Sam joined her. As they felt for a pulse, they follow his outstretched hand and their eyes were drawn upward, to a display on the walls. At first they stared uncomprehendingly at the neat, closed containers of seeds. Sam jumped up and examined them more closely.

The sign on the display read: GENETICALLY DIVERSE SEED.

A large headline from The Dallas Morning News was simply framed: SEED WARS THREATEN WORLD FOOD CROPS.

"They were here all the time, hidden in plain sight," he said.

Colby reached into the containers and almost reverently touched the seed. "Corn. Okra. Squash. Beans. Wheat. All indigenous seed. All safe seed."

Sam grimaced. "Even spinach. It's been in this school all these years, long before the blight."

Her voice was soft as rain. "Dad knew that."

"But he didn't tell them in time."

She nodded her head in agreement. There was no answer to that. Even if Dr. Ryder had told them the truth, would they still have killed him? Probably. She hesitated as if waiting for her father to forgive the past. "Now what?"

He smiled a little.

"There are plows in every barn in the county," he said.

"It will soon be April," she whispered.

They held each other, practically exploding with the possibilities.

———————

Involved as Sam and Colby were in their discovery, they did not hear closing around them the shuffling sound of someone coming. They turned to see two men, Angel and Dave. They hardly recognized them, they had changed so dramatically.

Neither Colby nor Sam felt any menace from them. What could the two old men do? They were near death. Still Colby felt as if her nightmare had returned.

Sam stepped out to meet them, unafraid, as if he knew them.

Angel spoke. "You know who we are."

"Yes. Now I think I do," Sam said. He sounded more certain than he was.

Angel's voice was weak, resigned. "What did you find?"

Sam pointed to the packets. "Uncontaminated seed."

"You have time then."

"Yes, spring will come soon," Colby said.

Angel stared at Potter and then at Sam. They began picking up Potter. The older man turned to Sam. "You didn't know him, did you?"

Sam looked puzzled. He pointed toward Potter. "Him? No, I guess not."

Sam stared at the man on the ground, and then, like a series of photographs clicking rapidly across a screen, he remembered. A snaggle-toothed boy. A fishing pole. A freckled-faced little guy with a fish. Sam reached into his pocket and pulled out the battered photograph from long ago. There was his young brother, holding the fishing pole and so proudly, the fish.

Sam's face reflected his anguish. "When I left, the last time I saw him, he was six. My baby brother. I read to him, played with him, but his name was John, John Mickelson," Sam said.

"He took the name of Potter from a child's book. The author's name. I don't know why."

The decrepit little man Angel shrugged. "Doesn't matter now."

Colby stepped up behind Sam and touched him. "You couldn't have known. No one could have known."

Struggling, the men lifted Potter. Sam helped carry him out to the street and placed him on the small trailer attached to the ATV. They formed a small funeral cortege and pulled him to the cemetery. Then they went out to Sam's home place and wrapped Papa Mickelson in quilts and brought him to be with his youngest son. Together, with difficulty and with hard work, by sunset they managed to bury Sam's father alongside the man who was but a boy.

Seventy Eight

Here The Heart Is

They were left alone then, Sam and Colby. Once Potter was buried, there was nothing more to say, about who he was and why they did not realize it for such a long time.

Questions were answered, and their work was cut out for them. They didn't want to dwell on what would be incredible difficulties that lay ahead. They began their tasks quickly.

There were tools and tractors and ethanol for all that they needed to accomplish. Jake had dropped everything he could spare to help them, and hoped that it would suffice. They were confident it was enough. Jake had made certain they had the things needed, and then, following his own orders, the crew resumed their mission to Mars. He had to be with those going toward the stars. He owed it to Eliza and to Colby.

Somehow, despite a heartfelt sadness at being left behind, Colby didn't mind that she wasn't going. She believed that she had a more important mission here with Sam.

By the middle of spring, they had planted the original seeds in the earth, and soon, the precious, tender green sprouts began. They heard from the Program Six team who had been in Colorado, and would soon begin their planting. They promised to remain in touch and at some point, they would hope to see one another.

Early that summer, she reached out her hand and gently touched the soft texture of the corn. It would ripen in a few weeks and they would harvest it, and life would be sustained. At some point they would figure out a way to go to Norway. She was a pilot after all, as was Sam.

She felt a presence behind her and turned. It was Sam, looking at her, smiling at her. "We've done it," he said.

She nodded her agreement.

In the evening, they opened the doors and windows of the little church, the light a soft backdrop for dust motes floating in lazy circles. The early summer breeze cooled the building and dissipated the smell of mildew. They had decided that there would be traditions in their new world. Together they cleaned the dust filled building, and in the Sunday morning following, they knelt together at the altar and took turns reading the sacred vows of marriage.

Somewhere, on the hill in the distance, bright with the sun and green growth of grass and trees, two of the survivors, Angel and Dave, watched through binoculars, but soon set them down. They were too heavy to hold any longer. They had made a little camp up there with canned goods, sleeping bags and a tent. They would not leave that place.

Jake had given up the idea of returning to earth. Instead, like Noah, he would shepherd his young team farther into the universe, until the end of his days. But not before a chosen few were sent to join Sam and Colby.

Far out in the dark cold expanse of endless sky, the inhabitants of Polaris made preparations to send the next convoy home. The massive sky-built edifice was a city in the stars. They would be bringing new growth of plants and trees and creatures toward the turning earth and gradually in their lifetimes they would see the green flow over and blank out the deadly browns of death. A group of survivors living on the rooftops of Manhattan and Brooklyn managed to send a brief message with a promise for further connection. In Norway, there was excitement in the Cavern. Four other young couples would be joining them soon. Perhaps there were more.

– The End –

AFTERWARD

What can you say about the courage of those who began life anew? How can you make any judgment about the reality of experiencing total loss, and picking up the pieces to forge a new beginning? It has been done many times in human history, and it will be done again. My great grandparents were pilgrims of a sort, deep in the heart of Central Texas in 2043 and for many years following. They worked tirelessly there, and then journeyed to meet other survivors of the great Seed Wars. They ultimately prevailed, which explains my presence. My grandfather Samuel told me that I must never forget the lessons of the disaster that decimated the earth. I promised him this some years ago, and today I keep the promise. The earth offers many gifts, and we must not take them lightly. To do so is to invite catastrophe.

I pledge to protect the cave, the seeds, the fish and bees, the best of nature joined with the careful additions of man. I do not fancy another Life Hunt. Too much is at stake, wouldn't you agree?

Samuel III

Bibliography

"A Tipping Point for GM Foods?" *The Wilson Quarterly*, Spring 2008.

Abbott, Charles. "Biotech Food Rules Likely to Get Tighter." Reuters. com. http://reuters.com (accessed August 17, 2000).

Allman, William F. "Pesticides: An Unhealthy Dependence?." *Science*, October 1985.

"Amaranths." *1995 Seedlisting*, Summer 1995.

Amaya, Nan. "Progess on CFC Emissions." *Lone Star Sierran (Austin)*, December 1, 1989.

Ambrose, Sue Goetinck. "Tough row to hoe." *The Dallas Morning News*, April 10, 2000, see. Discoveries.

Associated Press. "New family of genes reportedly may help fight off crop disease." *The Dallas Morning News*, September 26, 1994.

Bilger, Burkhard. Annals of Gastronomy "True Grits. A chef's quest for authentic Southern food." The New Yorker. October 31, 2011. p. 40.

Bishop, Jerry E. "Scientists Report Inserting Gene into Corn Plants That Stay Fertile." *The Wall Street Journal (New York)*, January 24, 1990, sec. Technology and Science.

Borenstein, Seth and Malcolm Ritter. "Is That the Real thing on Your Plate? The Science of Altered Food." The Associated Press. Sunday, October 3, 2010. p. 34A. *The Dallas Morning News*, "Chips pulled over corn concerns," July 5, 2001.

Choi, Charles Q. "Design Revealed for Doomsday Seed Vault." LiveScience | Science, Technology, Health & Environmental News. http://livescience.com (accessed February 9, 2007).

Chotzinoff, Robin. "Edible Heirlooms." *Food & Wine*, August 1996. *The Washington Post*, "Concern about altered grain closes Kellogg plant," October 21, 2000.

Crenson, Matt. "A Blizzard of Action." *The Dallas Morning News*, November 1, 1993, Discoveries section.

"Decline in genetic diversity spells disaster." *Utne Reader*, Sep. - Oct. 1988.

Drew, Lisa. "The Barnyard Restoration." *Newsweek*, May 29, 1989. "Fiddling With Food." CBS News. http://cbs.aol.com (accessed January 27, 2000).

"Genetically engineered foods causing some concerns." CNN.com. http://www.cnn.com (accessed September 27, 2000).

Harney, Thomas. "The Ozark wild gourd: ancestor of summer squashes." *Smithsonian Institute Research Reports* 70 (1992): 1,6.

Hart, William. "Center saving seeds of native farming." *The Dallas Morning News*, February 23, 1986.

Howard, Judy. "Plano projects chosen." *The Dallas Morning News*, May 15, 1985.

The Huffington Post. "Vietnamese Woman 'Ages' 50 years in Days. (October 14, 2011)

"International Space Station." Wikipedia. http://wikipedia.com (accessed June 14, 2010).

Jaret, Peter. "The good, the bad and the genetically engineered." CNN.com. http://www.cnn.com (accessed September 27, 2000).

Kemp, Jack . "Bioengineering lets farmers feed the world." *The Dallas Morning News*, December 9, 1999.

Kilman, Scott. "Some of Midwest Corn Crop is Poisoned With a Carcinogen, Inspectors Discover." *The Wall Street Journal (New York)*, September 26, 1988, sec. Commodities.

Kilman, Scott. "King of Corn Has Tips For Farmers, but More Isn't What They Need." *The Wall Street Journal (New York)*, February 28, 2000.

Kilman, Scott. "Bioengineered Bugs Stir Scientific Dreams But Will They Fly?" *The Wall Street Journal (New York)*, January 26, 2001.

Kirlin, Katherine S, and Thomas M Kirlin. "Off The Shelf." *Smithsonian Institution Research Reports*, Summer 1991.

Lee, Steven H.. "Texas farmers withholding judgment on modified crops." *The Dallas Morning News*, December 25, 1999, sec. Business.

Lindsay-Herrera, Flora. "Bad Seeds." *The Wilson Quarterly*, Spring 2008.

MacDougall, Ian. "Norway doomsday seed vault hits 1/2 million mark." RR.com. http://www.rr.com/news (accessed March 13, 2010).

Mellgren, Doug. "'Doomsday' Seed Vault Opens in Arctic." RR.com. http://www.rr.com (accessed February 26, 2008).

Mitchell, Garry. "Insects Defy Resistant Cotton." The Associated Press. http://ap.org (accessed July 24, 1996).

Morris, Tim. "Eating Our Words." *The Wilson Quarterly*, Spring 2008.

Myerson, Allen R.. "Breeding Seeds of Discontent." *The New York Times*, November 19, 1997, sec. Business Day.

Peed, Mike. "We Have No Bananas. The New Yorker. January 10, 2011. p. 28.

Ponte, Lowell. "SOS- Save Our Seeds!" *Readers Digest*, June 1987.

Powledge, Tabitha M. "Gene Pharming." *Technology Review*, Aug. - Sep. 1992.

"Regulated or Not, Nano-Foods Coming to a Store Near You." Aolnews. com. http://www.aolnews.com (accessed March 24, 2010).

Rubin, Rita, and Steven H. Lee. "Tainted feed, milk discarded." *The Dallas Morning News*, October 1, 1988.

Ryan, Missy. "GMO Rice Safe Without Oversight, USDA Says." Reuters. com. http://reuters.com (accessed November 25, 1999).

"Scientists Eye Potato Disease." The Associated Press. http://ap.org (accessed June 20, 1996).

Seabrook, John. "Tremors in the Hothouse." *The New Yorker*, July 19, 1993.

Shell, Ellen Ruppel. "Seeds in the bank could stave off disaster on the farm." *Smithsonian*, January 1990.

"Soil-ree Crops Good for Wheat." The Associated Press. http://ap.org (accessed July 2, 1996).

Sokolov, Raymond. "Subterranean Treasures." *Natural History*, January 1991.

Sokolov, Raymond. "One Man Bites Back." *Natural History*, July 1991.

Specter, Michael. "The Pharmageddon Riddle." *The New Yorker*, April 10, 2000.

Steimel, Dirck. "Protecting tomorrow's crops." *The Dallas Morning News*, July 3, 1988.

"Study calls for more regulation, not labeling, of biotech crops." CNN. com. http://www.cnn.com (accessed September 27, 2000).

"Study: Gene-modified corn releases insecticide in soil." CNN.com. http://www.cnn.com (accessed December 3, 1999).

Tangley, Laura. "Seed savers get a break." *U.S. News & World Report*, October 18, 1999.

Terdiman, Daniel. "America's Fortress: Cheyenne Mountain is alive and kicking." Technology News - CNET News. http://news.cnet.com (accessed June 27, 2009).

Winslow, Ron. "'Fungus Fatale' Poses a Threat to Potato Crop." *The Wall Street Journal (New York)*, January 18, 1995.

Witze, Alexandra. "Scientists work at genetic level in effort to build a better tomato." *The Dallas Morning News*, June 27, 1994, sec. Discoveries.

Yoon, Carol Kaesuk. "Genetic contamination found in native Mexico corn varieties." *The Dallas Morning News*, October 15, 2001.

Johnson, Kirk, Friday, October 8, 2010. The New York Times, Health and Science. p. 14A. The National Geographic. "How Heirloom Seeds Can Feed the World." July 2011. p. 108.

Note to Reader: These are actual published stories relating to the topic of genetic engineering of food. They were collected over a number of years.

About the writers: Neila Skinner Petrick is an award winning author who specializes in the Southwest. Barry Chambers is author of the Rattler series, Willy the Hit Man and more than a dozen produced plays. He is an actor as well. Ivy Opdyke is a multi-talented actor who has appeared in film and on stage..

www.ingramcontent.com/pod-product-compliance
Lightning Source LLC
Chambersburg PA
CBHW051442170526
45166CB00001B/78